The Natural Limits to Biological Change

About the Authors and Respondent

LANE P. LESTER Lane P. Lester is Professor of Biology at Liberty Baptist College in Lynchburg, Virginia, and has previously served on the faculties of the University of Tennessee at Chattanooga and Christian Heritage College. He is a graduate of the University of Florida (B.S.E., biology) and Purdue University (M.S., ecology; Ph.D., genetics). Dr. Lester has co-authored three books, including *Cloning: Miracle or Menace?* and *Student Experiments in Genetics.* His numerous articles have appeared in *The Virginia Journal of Science, American Biology Teacher, The Science Teacher, Creation Research Society Quarterly,* and *Moody Monthly.*

RAYMOND G. BOHLIN Raymond G. Bohlin is currently Research Projects Manager at Probe Ministries International, Dallas, Texas. He is a graduate of the University of Illinois (B.S., zoology) and North Texas State University (M.S., population genetics), and is currently a student in the doctoral program (molecular biology) at the University of Texas, Dallas. Since 1979 Mr. Bohlin has been a guest lecturer on more than two dozen college and university campuses. He addresses issues in the creation-evolution debate and related scientific topics, especially genetics. His articles have appeared in the *Journal of the American Scientific Affiliation, Journal of Thermal Biology, Journal of Mammology, Creation Research Society Quarterly,* and *Christianity Today.*

V. ELVING ANDERSON V. Elving Anderson is Professor of Genetics and Cell Biology at the University of Minnesota and Director of the Dight Institute for Human Genetics in Minneapolis. He is the author of numerous journal articles and several books, including *Christianity and Natural Science.*

The Natural Limits to Biological Change

Lane P. Lester
and
Raymond G. Bohlin

with a response by
V. Elving Anderson

ZONDERVAN PUBLISHING HOUSE
OF THE ZONDERVAN CORPORATION
GRAND RAPIDS, MICHIGAN 49506

PROBE MINISTRIES
INTERNATIONAL
DALLAS, TEXAS 75204

Copyright   Copyright © 1984 by Probe Ministries International

Library of Congress Cataloging in Publication Data

Lester, Lane P.
  The natural limits to biological change.
  Bibliography: p.
  1. Evolution. 2. Genetics. 3. Variation (Biology) 4. Creationism. I. Bohlin, Raymond G. II. Anderson, V. Elving (Victor Elving), 1921–   . III. Title.
QH371.L47   1984   575   84-7295

ISBN   0-310-44511-6

Place of Printing   *Printed in the United States of America*

Series Editor   Steven W. Webb, Probe Ministries

Design   Inside cover design by Paul Lewis
Book design by Louise Bauer

87 88 89 90 91 / CH / 7 6 5 4 3

What Is Probe?

Probe Ministries is a nonprofit corporation organized to provide perspective on the integration of the academic disciplines and historic Christianity. The members and associates of the Probe team are actively engaged in research as well as lecturing and interacting in thousands of university classrooms throughout the United States and Canada on topics and issues vital to the university student.

Christian Free University books should be ordered from Zondervan Publishing House (in the United Kingdom from the Paternoster Press), but further information about Probe's materials and ministries may be obtained by writing to Probe Ministries International, P.O. Box 801046, Dallas, Texas 75204.

Contents

Illustrations

Book Abstract

Evolutionary theories of biological change have dominated the life sciences for over a century. The authors of this book present evidence to show that the two current theories, Neo-Darwinism and punctuated equilibrium, are not without serious flaws. A third alternative, based on the overall interpretation that there may be limits to biological change, is presented as a rational and competing theory of biological change.

Darwinism or Neo-Darwinism has long emphasized the gradualness and slowness of the pace of evolution. The view of both Darwinism and Neo-Darwinism is that organisms are able to respond only to minor environmental fluctuations. Dramatic climatic and genetic alterations will only lead to extinction. This view has become increasingly criticized, not only in terms of the mechanism of mutation and natural selection but also for the lack of evidence for gradualism in the fossil record.

Those who hold to punctuated equilibrium have exploited the shakiness of Neo-Darwinism's foundation to emphasize their view that evolution occurred with a jerky and erratic pace. They declare that selection acts on the species as well as on the individual. But their argument may be more semantic than substantive, and there is no alternative offered for the origin of new genetic information.

The theory that there may be definable limits to biological change is an abrupt shift from evolutionary theories. It is an old idea that deserves fresh consideration based on evidence from a number of sources. That organisms respond to environmental fluctuations through genetic variations and natural selection is undeniable. However, this process will usually reach a certain point and then go no further because the genetic variability has been exhausted. One then faces the question of the origin of genetic variability and its replenishment. This is the focus of this book.

chapter one

The Nature of Biological Change

The problem of the nature of biological change is introduced. What are the limits, if any, to biological change?

That populations of living organisms may change in their anatomy, physiology, genetic structure, etc., over a period of time is beyond question. What remains elusive is the answer to the question, How much change is possible, and by what genetic mechanism will these changes take place? Plant and animal breeders can marshal an impressive array of examples to demonstrate the extent to which living systems can be altered. But when a breeder begins with a dog, he ends up with a dog—a rather strange looking one perhaps, but a dog nonetheless. A fruit fly remains a fruit fly; a rose, a rose, and so

13

on. So biological change is occurring, but just how far will nature go?

Mutations may produce no effect, little effect, or a drastic effect. But most, if not all, drastic effects are lethal or deleterious. Are minor mutations that have been accumulated over thousands of generations and millions of years capable of producing complex invertebrates where once there were only single-celled organisms? Or human beings where once there were only arboreal monkeylike creatures? Or do mutations in fruit flies, as in breeding experiments, produce only a different kind of fruit fly?

As new information is gathered about the mazelike complexity of the genetic machinery and the intricate, yet delicate nature of biological adaptation, the more the traditional evolutionary explanation is questioned. Neo-Darwinism, though still satisfactory and even compelling to many, is falling into disrepute. Punctuated equilibrium on the other hand seems to be facing equally difficult challenges in establishing itself as a viable theory.

We believe that the available evidence can also lead to the conclusion that there are limits to biological change. The case presented in this book should lead, if not to the same conclusion, at least to the in-depth consideration of a scientific theory of limited biological change.

A point that we intend to elaborate on is the effect of presuppositions and a world view on this particular issue. Although there are various theories of biological change, the actual facts of nature do not change. What does change is the interpretation that can often be traced to the presuppositional differences among rival camps. This kind of scrutiny is critical because certain presuppositions may be held to so tightly that fallacies in the theory in crucial areas can be missed. It often requires a different presuppositional viewpoint for one to perceive the problem.

chapter two

The Variety and Complexity of Life

The biological world exhibits a diversity of adaptation. It also reveals a remarkable beauty and variety of form. In this chapter we give an overview of the extent of variation and complexity of adaptation to gain a perspective on what must be explained by any theory of biological change and origin.

The H.M.S. *Beagle*, on which Darwin sailed, was by no means the first ship to visit the Galapagos Islands. Western man first came to the islands in the early sixteenth century. The early visitors were most fascinated by the giant tortoises for which the islands were subsequently named. These great reptiles, sometimes weighing up to five hundred pounds and having shells as big as bathtubs, have no natural predators or competitors (Figure 1). On this curious archipelago the tortoises thrive and often live for well over one hundred years. But   15

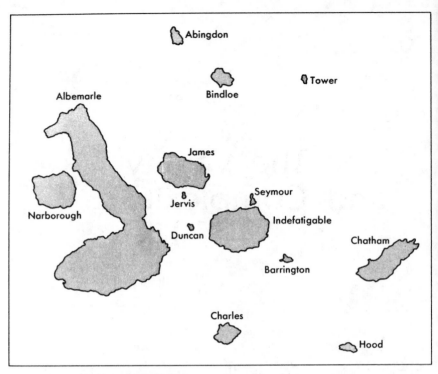

Duncan

Figure 1. Carapace Variations of Galapagos Tortoises. Drawings show the characteristics of three different species of tortoises found on different islands in the Galapagos. The longer-necked species live in relatively dry places and feed on tree cacti, the species with the short, straight neck lives in moister regions and feeds on dense, low-growing vegetation. Source: G. Ledyard Stebbins, Processes of Organic Evolution (P.O.E.) (Englewood Cliffs, N.J.: Prentice-Hall, 1971), p. 6. Used by permission.

Abingdon

Albemarle

apart from their renown as a curious work of nature and as a source of fresh meat for sailors on long journeys, these patriarchs of the archipelago received little notice until the arrival of Charles Darwin in 1835.

Darwin too was fascinated by the Galapagos tortoises. He clocked their "speed" (360 yards per hour) and spent hours observing them approach their water holes and take long drinks with outstretched necks and submerged heads. However, little would have been added to our knowledge of these beasts had Darwin not discussed them with the acting British governor of the Galapagos Islands, Nicholas Lawson.

Variation

Darwin was astonished when Lawson told him that he could identify which island a tortoise came from by the shape of its shell (carapace). Not only were there differences in the shape of the carapaces but also less noticeable differences in their size, color, and thickness. This kind of interisland variation was a totally new and unexpected concept to the young naturalist. Over a century later, the tortoises, though close to extinction, are classified as a single species with fifteen subspecies. Ten of these fifteen are on ten separate islands. The remaining five are isolated on the five principal mountains of Albemarle Island. Their primary distinguishing characteristic remains the size and shape of the carapace.

Darwin's learning of the interisland variations among the tortoises only added to his curiosity. Earlier he had begun to notice variations in the size of the beaks of a group of finches on the islands. Although the variations in characteristics among the finches is not totally analogous to those among the tortoises, the latter added to the impact on Darwin. He began cataloguing his specimens according to individual islands rather than lumping them together. But even so, the importance of the variation that Darwin recognized would not be fully realized by him or by anyone else for many years after the publication of the *Origin of Species*.

Of the twenty-six species of birds occurring naturally on the Galapagos Islands, thirteen of them are of finches. These finches are obviously related to each other on first impression. They range from a drab gray to a brownish color. In a few species the males possess a totally black dorsal plumage (Figure 2). Although remarkably similar in some characteristics, they are just as remarkably different in others. They differ primarily in what they eat, how they get their food, and where they find it. One group of finches of the genus *Geospiza* consists of ground foragers. Each species has a slightly different-sized beak and consequently concentrates on a different-sized seed. A second group of finches (genus *Camarhynchus*) forages mainly in the trees. These finches, except for one vegetarian species, feed primarily on insects of all sizes. The "woodpecker finch" is among the second group. It uses cactus spikes to probe cracks and crevices of dead trees for insects. Another finch (*Certhidea olivacia*) fills the role of a warbler. This warbler finch has the characteristic size and shape of a warbler and occurs in almost all of the islands and habitats.

The finches of the Galapagos Islands have come to be known as Darwin's finches. Anywhere from three to ten species occur on any one island in different combinations. As with the tortoises, the extent and pattern of variation among these finches greatly influenced later investigators in their evaluation of the nature and function of variation within (intraspecific) and between (interspecific) species. The extent and utility of variation within natural populations has remained a primary pursuit of biology to the present day.

Marked visible variation is pervasive in the natural world. The presence of distinct morphological forms within a species or population is

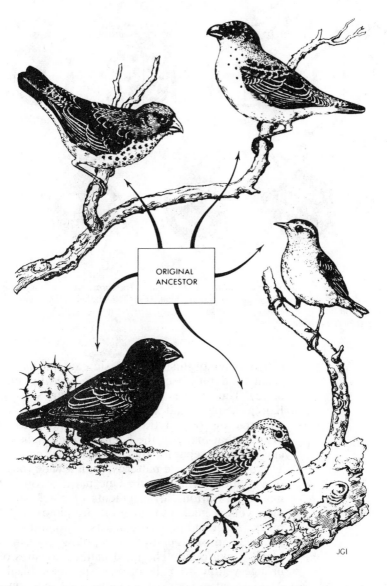

Figure 2. Darwin's Finches. Some of the different species of Galapagos Finches, which originated from a common ancestor through adaptive radiation in association with the different habitats available on the Galapagos Islands. From Lack, Darwin's Finches (Oxford University Press). Taken from Stebbins, P.O.E., p. 151. Used by permission.

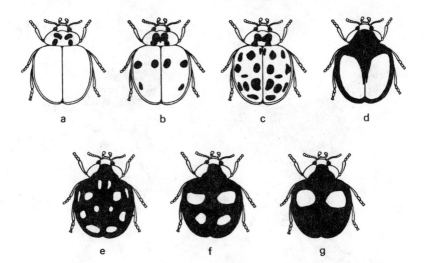

Figure 3. Variant Color Patterns in the Asiatic Beatle. The forms are: a-c, succinea; d, aulica; e, axyridis; f, spectabilis; g, conspicua. From T. Dobzhansky, Genetics of the Evolutionary Process (New York: Columbia University Press, 1970), p. 270. Used by permission.

called polymorphism. A classic example is the Asiatic ladybird beetle, *Harmonia axyridis* (Figure 3). This species is found throughout Siberia, Korea, Japan, and China. The variant color patterns on the elytra (wing covers) are so distinct that many of the beetles were originally classified as separate species and even genera. There are four basic patterns, with a wide range of variations within each. One pattern consists mainly of various arrangements of black spots on a yellow field, though some are almost solid black. A second pattern consists of a number of arrangements of a pair of large yellow spots on a black background. The third pattern consists of orange-yellow to pale orange spots on black. The last pattern consists mainly of red spots on a black background. All of these variants are not only distinct—they are also discrete. Although the various groups freely interbreed, there are no intermediates.

Another good example of polymorphism is the peppered moth of England. Although this moth is better known as an example of natural selection (which will be discussed later), it is also a striking example of variation within a species. There are only two forms of this moth, a melanic or dark form and a peppered or light form. The melanic form is prevalent in areas where the trees are devoid of lichens because of pollution and hence the tree surface is dark. The peppered form persists in areas where the lichens remain, giving the tree surface a lighter and mottled appearance. Such examples of distinct polymorphisms are the obvious, yet rare, cases of variation in nature. In most cases, variation is less striking. Usually there is less contrast and only a trained eye will notice the difference. One need only think of the variety of colors of roses, the many breeds of dogs, and the often confusing array of human facial features to see that the differences can often be very subtle.

Biological organisms, however, are not fascinating only to the degree to which they are similar yet dissimilar. Each group of organisms possesses a unique set of capabilities that allows it to live harmoniously within its particular environment. These capabilities are called adaptations.

Adaptations

Ranging from the expected to the bizarre, from the unimaginative to the incredible, adaptations are a large part of what makes biology fascinating. The deep-sea anglers, for example, are a group of fish that covers one end of this spectrum. They appear to belong to science-fiction rather than the real world. Inhabiting the deep oceans at depths of over one mile, the deep-sea anglers possess some unusual adaptations.

The deep-sea anglers belong to a group of marine fish (order Laphiiformes, suborder Ceratioidei) whose foremost dorsal spine in the female is located on the head and elongated into a "fishing rod" tipped with a "fleshy" bait (Figure 4). The bait is dangled in front of her mouth and when another fish or invertebrate comes near enough to investigate a possible meal, the curious one suddenly becomes a meal. However, at a depth of one mile, there is no light; how, then, is the bait seen? Most deep-sea fish have some kind of light-producing capability. In the angler, it's contained in the bait. The light is believed to be what actually attracts the prey. The light is produced by the oxidation of luciferin by molecular oxygen with the aid of the enzyme luciferase. Nearly 100 percent of the energy produced by this reaction is in the form of light. A common household lightbulb, by contrast, transforms most of its energy into heat, not light.

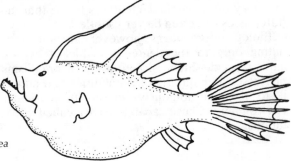

Figure 4. Deep-Sea Angler Fish.

But the deep-sea angler possesses one other unique feature. A specimen found in 1922 had a number of smaller fish attached to its abdomen. Initially these appeared to be its young. But later discoveries revealed that these "parasites" were actually the male of the species. Upon emergence from the larval to the adult state, the male searches out a female and bites into her abdomen. Eventually their tissues blend and the circulatory systems unite into one and the male literally lives the rest of his life as a parasite of the female. The all-important matter of finding a mate in a pitch-black sea is thus uniquely solved. Also, once the female's eggs are fertilized, it is believed that they rise to the surface in a jellylike mass. As the young fry emerge and mature, they eventually find their way down to the pitch-black depths of their ocean home.

An adaptation on the more prosaic side is the lack of a swim bladder in this fish. In most fish there is a small air sac that provides sufficient buoyancy to prevent sinking. But sinking to the bottom is precisely what the deep-sea angler requires; hence, it has no swim bladder. Besides, even if one were present, it would not be able to withstand the tremendous pressures of the deep ocean (at these depths the pressure is in excess of two thousand pounds per square inch). We see, then, that the deep-sea angler's body is adapted to suit its environment.

The Woodpecker's Adaptation

For another example of adaptation, let's move from the black depths of the ocean to the canopies of the forest. Almost any forest or grove of trees will do, for we are in search of the woodpecker, a common yet unique bird found in many parts of the world. This amazing bird possesses a set of adaptations that allow it to hammer its beak into hardwood trees and not only survive but actually use this extraordinary ability as its primary means of survival. The sound of a drumming woodpecker has always been intriguing. Even tapping our fingers in time to a woodpecker's tapping sound is difficult enough, but the thought of keeping up that pace with head movements is staggering. It is clear that the woodpecker possesses an exceptional set of muscles and nerves in its head and neck. This intricate network sustains amazingly precise and rapid movement, something rare for a vertebrate.

The woodpecker combines this hammering capability with four other structural features that enable it to go about its business of probing tree bark and chiseling nest holes. A woodpecker spends most of its time on vertical tree trunks; rarely is one seen either on the ground or perching. One look at the arrangement of toes on a woodpecker's feet explains why this is so: each foot has two claws in front and two in back. Most perching, grasping, and walking birds have three toes in front and one in back. This "two-plus-two" toe pattern is a major diagnostic feature for the entire order, Piciformes, to which the woodpecker belongs. This feature, along with stiff yet elastic tail feathers, allows a woodpecker to grasp a tree firmly and balance itself on a vertical surface. When the woodpecker braces itself to chisel a hole, the tail feathers bend and spread, buttressing the bird against the rough tree surface. In this way feet and tail form an effective tripod to stabilize the blows of hammering into wood.

An Amazing Beak and Tongue

The beak of a woodpecker is exceedingly tough, penetrating wood that could sometimes bend a nail. Between the beak and the skull is a piece of cartilage that acts as a shock absorber for the skull. The skull itself is thicker than that of most birds to help take the pressure of continuous pounding. But the tongue of a woodpecker is in a class by itself. When chiseling into a tree, the woodpecker will occasionally come across insect tunnels. Its tongue is long and slender and is used to probe these tunnels for insects. The tip is like a spearhead with a number of barbs or hairs pointing rearward. This facilitates securing the insect while transporting it to the beak. A sticky gluelike substance coats the tongue to aid in this process as well. But what is most amazing is the morphology of the tongue. Most birds cannot extend their tongue the full length of their beak, but a woodpecker stretches its tongue *three* to *five* times the length of its beak. The tongue is supported from within (as in all birds) by a shaft of bone (urohyal) that

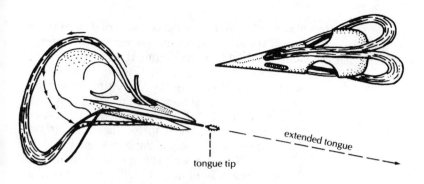

extended tongue

tongue tip

Figure 5. Green Woodpecker, Tongue and Morphology, After A. J. Marshall, ed., Biology and Comparative Physiology of Birds (New York: Academic, 1961), p. 64. Used by permission.

is driven from behind through a pair of slender bones (hyoid horns). In the woodpecker, these loop down into the throat, under the skull, around the back of the skull, beneath the skin, and over the top between the eyes, terminating usually just below the eye socket (Figure 5). In the European green woodpecker and the North American yellow-shafted flicker, the hyoid bones extend into the right nostril, and in the green woodpecker they extend the full length of the beak. When the green woodpecker extends its tongue, the muscles surrounding the bones contract, causing the horns to exit the nostril to a point directly between the eyes. The loop that extends into the throat flattens out to give full extension. The woodpecker indeed is no ordinary bird, yet it is a marvelous illustration of the principle of adaptation.

Amazing Larvae

Many other fascinating examples of biological adaptation are worthy of mention here. Recently, for example, K. Hagen of the University of California at Berkeley and J. Johnson of the University of Idaho released information about the amazing antics of the larvae of the beaded lacewing insect and a California termite. It was observed that lacewing larvae enter termite-in-fested branches and emerge some time later as adults. When this environment was duplicated in the laboratory, the larvae were observed to approach worker termites thirty times their size and wave their abdomens in the air. Each time, the target termite fell over completely paralyzed. The larvae then moved in for the meal. It was determined that when a larva waved its abdo-men, an allomone was released that paralyzed the termite. Allomones are chemicals that one species produce and that affect another species. It was also found that the allomone of the beaded lacewing affects only termites. Some have suggested that this discovery could turn out to be very practical if the allomone could be produced as an insecticide. This is of course a fascinating thought, but the point being made here is that the principle of biological adaptation is observable in many diverse ways in nature.

Even coloration plays a part in adaptation, and nature abounds in color. For many organ-isms color is not merely aesthetic in its charac-ter: it serves as a major tool of survival. Quite simply, the right coloration attracts or scares away, contrasts or blends with the surroundings, distinguishes or mimics. The sparkling colors of flowers attract the proper pollinators. The bright red and yellow shoulder patch of the male red-winged blackbird establishes rank among con-temporaries and helps attract females. The eye-spots on the wings of a moth can cause a would-be predator to pause a crucial instant. In the tropics there is a preponderance of green birds, toads, snakes, and insects in the trees, while on the forest floor such creatures come in various shades of brown. Some animals such as the ptarmigan, some northern grouse, and the arctic fox are able to change color to match the season. In Manila a white spider with yellow legs hides in white flowers with yellow stamens. The Malayan mantis hides in a special rhododendron of identical color. Contrasting black and white bands can cause the eye to momentarily lose all concept of form and see the pattern rather than the animal that wears it—e.g., the killdeer, the plover, and the angelfish. A certain spider in Java when at rest exactly resembles a bird's droppings.

Thus we have seen that whether life exists in the ocean, on the forest floor, or anywhere else, each group of organisms possesses a set of complex adaptations that uniquely equips it for survival. Mimicry is one broad category of adaptation that applies to many different species in a variety of environments. Let us consider some instances of mimicry.

Mimicry

Distasteful animals and plants that display bright warning coloration often are mimicked by palatable, unrelated forms as a protective maneuver. This was first discovered by Henry Bates in the nineteenth century while he was investigating the Amazon region of South America. In North and Central America, where it feeds on species of the milkweed *Asclepias*, the monarch butterfly is noxious to birds. The viceroy butterfly, a palatable species, would easily fool the amateur butterfly collector into thinking it was a monarch. Apparently the birds are fooled also, because they avoid the viceroy as well.

A more complicated example is found in some African butterflies. There are five distasteful models and five corresponding mimics. These models are five separate species representing three genera, but remarkably each is mimicked by a different variety of a single species, *Papilio dardanus*. Another form of mimicry is Mullerian mimicry. This type of mimicry involves the resemblance of two or more unpalatable organisms. Bees and wasps best display this form of mimicry in their alternating yellow and black bands. Even humans learn quickly to avoid this pattern.

Another interesting variation of mimicry is the proverbial wolf in sheep's clothing; that is, a predator disguised as a harmless coinhabitant. An example of this is the zone-tailed hawk of Central America, which resembles the turkey vulture in both appearance and soaring behavior. Since vultures are carrion eaters, small mammals pay them no attention. Zone-tailed hawks have been observed soaring with turkey vultures, and many a rabbit has undoubtedly fallen victim to a "turkey vulture" that unexpectedly came plummeting out of the sky.

These examples demonstrate that mimicry is a very effective tool of survival—one that has been used by numerous species.

Symbiosis

While the organisms involved in mimicry do not directly interact, symbiosis involves a close relationship between very different species and is another broad category of adaptation.

Symbiosis is best defined as an intimate, often obligatory interaction between two species that benefits at least one and sometimes both species. A common example is lichens, which are made up of algae and fungi. Algae provide organic compounds through photosynthesis, and the fungi dissolve nutrients from bare rock and bark. Root nodules of legumes are another example. Plants of the legume family—such as peanuts, peas, and alfalfa—nourish bacteria in their roots that in return fix nitrogen from air pockets in the soil. Termites and ruminant mammals such as cows and sheep have specialized bacteria as digestive aids, and they in turn provide the bacteria with a steady stream of nutrients.

The relationship between flowers and their animal pollinators has long been recognized as an intimate example of symbiosis. The plant-pollinator relationship is highly developed in the orchid family with its wide variety of shapes, colors, and fragrances. In Central America different species of orchids are highly specific in regard to the species of bee they attract. Since no nectar is produced by these orchids, the bees are attracted by the specific fragrance. Only males are attracted, females are indiscriminate. By using unique combinations of fragrances, orchids of the genus *Stanhopea* attract only one species of bee pollinator. The mechanism of pollination of this genus also aids in preventing cross-pollination. The bee *Eulaema meriana* enters the orchid *Stanhopea grandiflora* from the side (Figure 6). The lip of the flower is slippery and the bee sometimes slips as it withdraws from the flower and brushes against the column of the flower. The pollinaria then transfer to a particular point on the bee's body. If a bee with attached pollinaria slips and falls, the pollinaria may stick to the stigma, which is in a location that assures pollination for this species.

As an intriguing example of complex animal interactions in symbiosis, consider the mutually beneficial relationship between the greater African honey-guide (*Indicator indicator*), a bird, and the honey badger or ratel (*Mellivora capensis*). The bird makes its presence known by noisy chattering. The badger, taking notice, is then led to a beehive by the bird. The badger breaks open the hive and has its fill of honey and bee larvae. The honey-guide then moves in to gulp down fragments of wax, which it is able to digest. On its own, the honey-guide is unable to get at the wax of a bee hive. Similarly, the badger is unable to locate beehives. But together, they "serve" each other's purpose while satisfying themselves. As a nestling, the greater honey-guide illustrates a different, much less positive, symbiotic relationship. It is a brood parasite. The adult bird lays its eggs in the nests of other species such as barbets and woodpeckers. Upon hatching, the nestling will either eject its foster siblings from the nest or ferociously attack them, using the needle-sharp hook on its beak to bite and tear at its nestmates until they

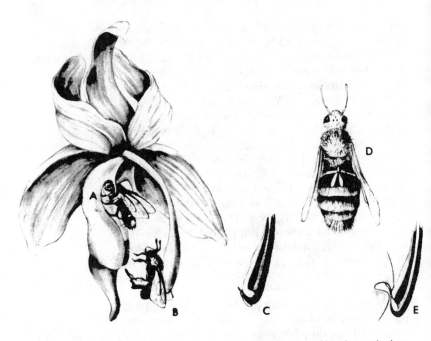

Figure 6. Orchid Pollinizations. The bee enters from the side and brushes at the base of the orchid lip (A). If it slips (B) the bee may fall against the pollinarium which is placed on the end of the column (C), and the pollinarium becomes stuck to the hind end of the thorax (D). If a bee with an attached pollinarium falls out of a flower the pollinarium may catch in the stigma (E), which is so placed on the column that the flower cannot be self-fertilized. From R. E. Ricklefs, Ecology (Newton, Mass.: Chiron, 1973), p. 199. Used by permission.

are dead. The nestling is then raised to adulthood as the "only child" of its foster parents. Thus, whether as an adult or as a nestling, the greater African honey-guide benefits mutually or parasitically from interactions with another species.

Cleaning Symbiosis   In the 1940s and early 50s, scuba divers began to bring back reports of fish hanging motionless while smaller fish poked around it, apparently feeding on parasites on the outer surfaces of the larger fish. Today this is commonly known as cleaning symbiosis. It is found most frequently in shallow tropical waters. Some marine fish and invertebrates are specialized in cleaning parasites and diseased tissue from fish many times larger than themselves. Some relationships are very casual, others exhibit a high degree of organization. One such intricate relationship exists between the Pederson shrimp (*Periclemenes pedersoni*) and its clients in the Bahamas. This tiny shrimp is usually found in association with the sea anemone *Bartholomea annulata*. As a fish approaches, the shrimp waves its antennae back and forth to attract attention. if interested, the fish swims over and allows the shrimp to climb aboard. The shrimp moves rapidly over the fish, picking off parasites, investigating irregularities, and cleaning injured areas. As the fish lies quietly, as if in a state of suspended animation, the shrimp is even allowed to make minor incisions to extradite subcutaneous parasites. Eventually, the shrimp is allowed full access to the gills and mouth cavity. All this is done with no attempt to harm the shrimp, which would no doubt make a tasty morsel. In no time at all the location of these shrimp becomes well known, and fish can be seen literally lining up to wait their turn.

The natural world is full of amazing variation and complex adaptation. The extent of visible variation and the complexity of adaptation is staggering. It is easy to see how one could find the study of living things vibrant and exciting. The array of examples in this chapter causes us to search even further for more exotic phenomena and more complete comprehension of that which we do not yet understand.

It is our intent to discuss further the search for the origin and meaning of variation and adaptation. This ultimately comes down to a question of genetics and the genetic structure of species. The nature of biological change is a subject with a stormy past; most experts predict that the future will be just as stormy. Advances in modern genetics have made this endeavor not only more intriguing but also more complicated. We will focus on these advances in the next chapter.

chapter three

The Wonders of Modern Genetics

The biologists of Darwin's time could not have imagined the complexity of inheritance of living systems. All theories of biological change must be consistent with the intricacies of modern genetics. This textbook-style examination will also serve as a glossary for later discussion.

Modern genetics is a field of unbounded complexity. No discussion of biological change can ignore it. In order to press our investigation to higher levels, it is mandatory that we examine genetics now because it will enhance the reader's understanding later on.

We will explore the field in a brief but academic manner. The main subjects of this interchange are: DNA and gene theory, chromosomes, mutation, and genetic variation. These topics are covered in textbook style because of limited space. Genetics is an exploding field to

33

such an extent that no one individual is capable of keeping up with its rapidly expanding horizons. The need to investigate the common thread running through all living systems is at the center of the search for life's past[1] as well as its future.[2] The abbreviation for deoxyribonucleic acid, DNA, is the genetic symbol probably the most recognized yet least understood by the modern public. What is it? What is it made of? What does it do? How does it work? How is it organized? Although the answer to these questions is not completely understood, what is known could take volumes to delineate completely. We will therefore attempt only to summarize the basics.

DNA and Gene Theory

DNA is a complex organic molecule that is now known to be the vehicle by which hereditary traits are passed on from generation to generation. Before Watson and Crick proposed their model of DNA structure in 1953, there was a great deal known about DNA from numerous experiments by various individuals. First, DNA was thought to be composed of two or more strands, with each strand composed of a linked chain of nucleotides. Second, there were four different nucleotides, each composed of a nitrogenous base, a sugar (deoxyribose), and a phosphate group . Of the four bases, cytosine and thymine are single-ring structures known as pyrimidines. Adenine and guanine are double—ring structures called purines (Figure 7). Third, nucleotides were linked together by the phosphate group of one nucleotide to the sugar of the next nucleotide (Figure 8). Fourth, the multiple strands were 22 angstroms in diameter, and there was a repeating unit of 34 angstroms in length (1 angstrom $= 10^{-10}$m). And finally, the sum of the two purine bases (A + G) equaled the sum of the pyrimidine bases (T + C).

Taking this information and other facts known at the time, Watson and Crick proposed a

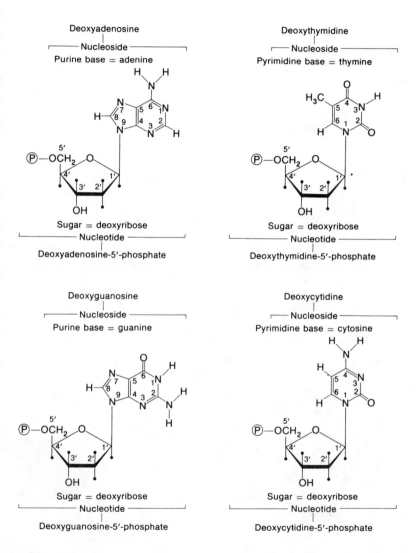

Figure 7. DNA Structure: Deoxyribonucleotides. From Geoffrey Zubay, Biochemistry (Reading, Mass.: Addison-Wesley, 1983), p. 662. Used by permission.

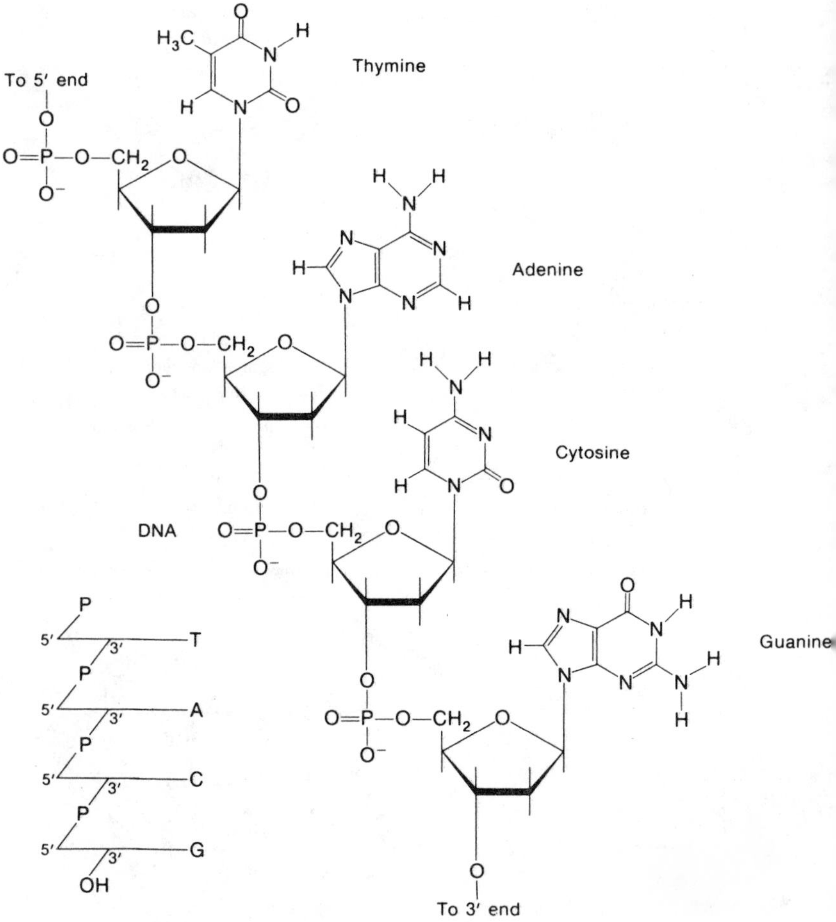

Figure 8. DNA Structure: Primary. In writing, the 5'-phosphate is indicated to the left and the 3'-phosphate is indicated to the right. The illustrated structure should be written as pTpApCpG. From Zubay, Biochemistry, p. 666. Used by permission.

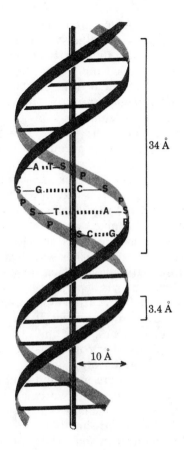

34 Å

3.4 Å

10 Å

Figure 9. DNA Structure: Watson and Crick Model. The molecule is composed of two polynucleotide chains held together by hydrogen bonds between their adjacent bases (S, sugar; P, phosphate; A, T, G, C, nitrogenous bases). The double-chained structure (shown here wound around a hypothetical rod) is coiled in a helix. The width of the molecule is 20 angstroms; the distance between adjacent nucleotides is 3.4 angstroms; and the length of one complete coil is 34 angstroms. From William Keeton, Biological Science, (New York: Norton, 1972), p. 508. Used by permission.

"double helix" structure that is now universally accepted. The DNA molecule consists of two strands of DNA that are coiled like a rope. The best way to visualize the structure of DNA is to think of a ladder that has been twisted (Figure 9). If you uncoiled the ladder and split it down the middle of each rung, each half would represent a single strand of DNA. The rail of each strand is composed of the alternating phosphate and sugar groups. Each half-rung represents a purine or pyrimidine base. The two bases that form the complete rung when the two halves are joined together are complementary. That is, a purine is always paired with a pyrimidine. More specifically, thymine is always paired with adenine, and cytosine is always paired with guanine. This arrangement assures that the sum of the purines will always equal the sum of the pyrimidines. Each rung of the ladder or each base pair is separated by 3.4 angstroms and is turned 36° from the preceding one. Therefore, a complete twist of 360° is achieved every ten base pairs. This unit of ten base pairs is therefore 34-angstroms long and represents the 34 angstrom repeating unit.

The base pairs are fitted and held together by hydrogen bonding. A hydrogen bond exists when a positively charged hydrogen atom is shared between two slightly negative atoms—in this case, a nitrogen and oxygen or two nitrogen atoms. Between adenine and thymine, there are two hydrogen bonds, and between cytosine and guanine, there are three. The length of these bonds combined with the overall width of each strand gives the molecule a diameter of approximately 22 angstroms as mentioned above. In order for these hydrogen bonds to be formed, it is necessary for the bases to be inverted in relation to each other, or antiparallel. This means that the opposing strand runs in the opposite direction or is upside down (Figure 10).

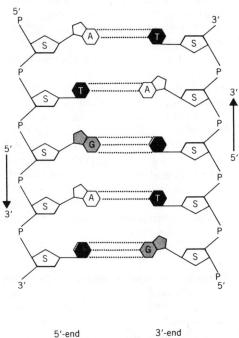

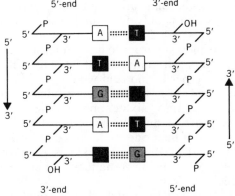

Figure 10. DNA Structure: Molecular Configuration and Opposite Polarity. From Gardner and Snustad, Principles of Genetics *(New York: Wiley, 1981), p. 81. Used by permission.*

DNA Replication

With the help of specific enzymes, the DNA molecule can be replicated precisely. The double helix is unwound, and the two strands are separated, exposing the bases along each strand. Using nucleotides that are in the form of high-energy triphosphates, each base along both strands is matched by its complement base. By this procedure, two identical DNA molecules are formed using each strand of the original as a template. Base complementarity, then, assures exact reproduction of the other strand. Adenine can be paired only with thymine, guanine only with cytosine, and vice versa. This potential for exact replication provides the basis for DNA to be the vehicle of transmission for inheritance. The genetic information is passed on accurately from generation to generation.

Protein Synthesis

The order in which nucleotides appear is important because it is the sequence of nucleotides that determines the sequence of amino acids in a protein. There are twenty common amino acids that are built into peptide sequences. With only four nucleotides available, nucleotides will have to be used in groups of three (triplets) in order to have at least one group code for each amino acid. Using only one nucleotide per amino acid would yield only four amino acids ($4^1 = 4$). Two nucleotides per amino acid yields only sixteen possibilities ($4^2 = 16$). Three provides an abundance ($4^3 = 64$), and this allows more than one triplet to code for a single amino acid. The genetic code is therefore referred to as being redundant or degenerate (Table 1).

The process by which the information contained in DNA is transformed into protein is complicated and involves a new set of molecules and enzymes. The principal new molecule is RNA or ribonucleic acid. RNA differs from DNA in two respects. The sugar within a nucleotide of RNA is ribose, which has an extra

FIRST RNA NUCLEOTIDE BASE	SECOND RNA NUCLEOTIDE BASE				THIRD RNA NUCLEOTIDE BASE
	U	C	A	G	
URACIL (U)	PHENYLALANINE	SERINE	TYROSINE	CYSTEINE	U
	PHENYLALANINE	SERINE	TYROSINE	CYSTEINE	C
	LEUCINE	SERINE	STOP	STOP	A
	LEUCINE	SERINE	STOP	TRYPTOPHAN	G
CYTOSINE (C)	LEUCINE	PROLINE	HISTIDINE	ARGININE	U
	LEUCINE	PROLINE	HISTIDINE	ARGININE	C
	LEUCINE	PROLINE	GLUTAMINE	ARGININE	A
	LEUCINE	PROLINE	GLUTAMINE	ARGININE	G
ADENINE (A)	ISOLEUCINE	THREONINE	ASPARAGINE	SERINE	U
	ISOLEUCINE	THREONINE	ASPARAGINE	SERINE	C
	ISOLEUCINE	THREONINE	LYSINE	ARGININE	A
	START/METHIONINE	THREONINE	LYSINE	ARGININE	G
GUANINE (G)	VALINE	ALANINE	ASPARTIC ACID	GLYCINE	U
	VALINE	ALANINE	ASPARTIC ACID	GLYCINE	C
	VALINE	ALANINE	GLUTAMIC ACID	GLYCINE	A
	VALINE	ALANINE	GLUTAMIC ACID	GLYCINE	G

NEUTRAL   AROMATIC   BASIC   ACIDIC   SULFUR-CONTAINING

TABLE 1

TABLE 1: DICTIONARY OF GENETIC CODE, tabulated here in the language of messenger RNA. The code is universal: all organisms, from the lowliest bacterium to man, use the same set of RNA codons to specify the same 20 amino acids. In addition AUG serves as a "start" codon to signal the beginning of the messenger-RNA transcript, and UAA, UAG and UGA serve as "stop" codons that signal the end of the transcript and cause the complete protein to be released from the ribosome. The code is highly redundant in that several codons specify the same amino acid. Nevertheless, certain point mutations (single substitutions of one nucleotide-base pair for another in the DNA molecule) may change a codon so that if specifies a different amino acid.

hydroxyl group located on the second carbon, as compared to deoxyribose in DNA. Also, RNA replaces the pyrimidine base thymine with uracil. Apart from these two differences, a single strand of RNA is constructed exactly like DNA. RNA is found in three distinct forms within the process of protein synthesis: first, as a single strand of messenger RNA (mRNA); second, in association with protein to form ribosomes (rRNA); and, third, as a specific adapter molecule known as transfer RNA (tRNA).

Protein synthesis proceeds in essentially two steps: transcription and translation. Transcription occurs in the nucleus, translation in the cytoplasm. It has long been recognized that DNA does not leave the nucleus. How then does a message get from the nucleus out to the cytoplasm? With the help of specific enzymes, DNA serves as a template from which mRNA is formed. This single strand of mRNA carries the information in its sequence of nucleotides to code for a single protein molecule. The mRNA leaves the nucleus and in eukaryotic cells (those with a distinct membrane-bound nucleus) locates the ribosomes in the endoplasmic reticulum (a system of membrane-bound channels in the cytoplasm). At the site of the ribosomes, with the help of a different set of enzymes, the protein molecule is assembled. Often a single mRNA is "read" by a number of ribosomes simultaneously. The resulting conglomerate is called a polysome. At the ribosome, tRNA molecules assemble together with the appropriate amino acid attached to each. The tRNA molecule contains within its nucleotide sequence the complement base sequence (anticodon) to the codon on the mRNA, the codon being the triplet of nucleotides as found on mRNA. It is by this process of recognition that the appropriate tRNA with the proper amino acid is placed into position on the protein molecule (Figure 11).

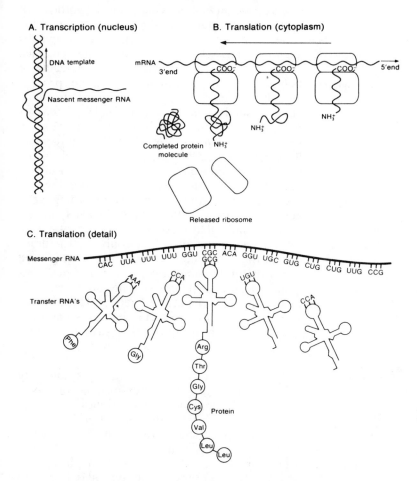

A. Transcription (nucleus)

DNA template

Nascent messenger RNA

B. Translation (cytoplasm)

mRNA
3'end
COO⁻ COO⁻ COO⁻
5'end

NH₃⁺ NH₃⁺

Completed protein NH₃⁺
molecule

Released ribosome

C. Translation (detail)

Messenger RNA
CAC UUA UUU UUU GGU CGC ACA GGU UGC GUG CUG CUG UUG CCG
                          GCG

Transfer RNA's

Phe
Gly
Arg
Thr
Gly
Cys
Val
Leu
Leu

Protein

Figure 11. Transcription and Translation. A. Transcription. One strand of the DNA helix serves as a template for the synthesis of a complementary chain of messenger RNA. B. Translation. In eukaryotes the messenger RNA synthesized in the nucleus moves to the cytoplasm, where several ribosomes attach to it. Each ribosome synthesizes a polypeptide as it proceeds several nucleotides behind the preceding ribosome. Each codon in the messenger RNA is recognized by a complementary anticodon on a transfer RNA molecule carrying a particular amino acid. C. Detail of the translation process. The configuration of the transfer RNA molecules is an outline of the known structure of alanine transfer RNA. From Dobzhansky et al. Evolution *(San Francisco: Freeman, 1976), p. 26. Used by permission.*

The Gene Concept

Since we have discussed protein synthesis, it is now appropriate to define the concept of the gene. The gene has been defined as a specific' segment of DNA that determines the sequence of amino acids for a single protein molecule or polypeptide. In accordance with the process described above, the cell transcribes then translates a single gene, with a single protein or polypeptide as the final product. This raises a new set of questions. How does the cell regulate which genes are transcribed? How much of the product is produced? And when does all of this take place? In other words, how does the cell regulate the various functions regarding gene action? To answer these questions, it is necessary to realize that there are two types of genes. So far, we have discussed only "structural" genes, those that code for a protein that is used as a structural component of the cell or performs some metabolic function. The other class is made up of "regulatory" genes. These genes code for protein products that in some way regulate the transcription of a structural gene.

Gene Regulation

The product of a regulator gene is generally called a repressor. The repressor molecule binds to the DNA, preventing transcription from taking place. In other words, the action is preventive or negative. The repressor may be active or inactive in its primal state. The action of a repressor can be compared to turning a light on or off. If the repressor is active, it is bound to the DNA and transcription is prevented; that is, the "light" is normally turned off. To turn the light on, another molecule, an inducer, is required. The inducer deactivates the repressor, rendering it unable to bind to the DNA, allowing transcription to take place unhindered. Inactive repressors perform an opposite function. The inactive repressor is unable to bind to DNA on its own. Transcription takes place freely. The light is normally on. The light is turned off when the repressor binds to a co-repressor that activates the repressor. The repressor then binds to the DNA, transcription is stopped, and the light is turned off.

These two regulatory systems can be characterized as an inducible system, involving an active repressor and a repressible system with an inactive repressor. Examples of both have been documented in living systems. An example of an inducible system is found in *E. coli,* which is the common bacteria that inhabit human intestines. Lactose is the precursor to an inducer, callo-lactose, which enables 6-galactosidase, the enzyme that metabolizes lactose, to be produced. This is the classic and most well-known regulatory system. A detailed discussion can be found in any modern introductory genetics text. Repressible systems are found in many amino acid synthesis pathways. Production of tryptophan is slowed by the accumulation of tryptophan in the cell. Tryptophan or a by-product acts as the co-repressor. When there is an abundance of tryptophan, the enzymes required for its synthesis are no longer produced.

Regulatory genes can also take on a positive role. If they do so, the regulator protein acts as an activator, and its role in either an inducible or repressible system is reversed (Table 2).

Regulatory genes are usually located in close proximity to the structural genes they regulate. There may also be other segments of DNA adjacent to the structural genes that function in the regulatory process. The stretch of DNA where the regulator protein binds is called the operator. The promoter region marks the side where the RNA polymerase, the enzyme that catalyzes transcription, attaches to begin the transcription process. It is important to note that the operator and promoter sequences are not genes, because they do not code for a specific polypeptide. This integrated unit of regulatory and structural genes, operator and promoter, is known as an operon. The structure and function

TABLE 2: REGULATION SYSTEMS COMPARISON

	Negative Control		Positive Control	
Regulator protein	Repressor		Activator	
Controlling site on DNA	Operator		Initiator	
Role of regulator protein	Prevents gene transcription		Enables gene transcription	
	Inducible System	**Repressible System**	**Inducible System**	**Repressible System**
Effector module	Inducer	Corepressor	Inducer	Corepressor
Role of effector	Prevents repressor function	Stimulates repressor function	Stimulates activator function	Prevents activator function
Presence of effector	Enables gene transcription	Prevents gene transcription	Enables gene transcription	Prevents gene transcription

Source: Strickbergher, Genetics, p. 689. Used by permission.

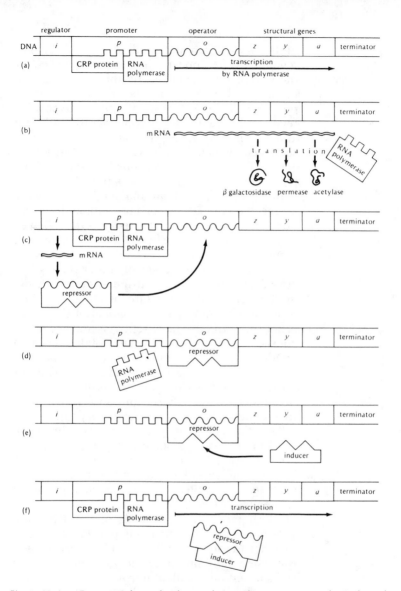

Figure 12. Lac Operon. Scheme for the regulation of Lac enzyme synthesis through transcription control (a) In the absence of repressor, transcription of the lac operon is accomplished by the RNA polymerase enzyme which attaches at the promoter site (p) in the presence of CPR protein, and then proceeds to transcribe the operator until it reaches the terminator sequence. (b) Subsequent translation of the lac mRNA leads to the formation of the lac enzymes. (c) Synthesis of the lac repressor protein occurs via translation of an mRNA which is separately transcribed at the i regulator gene locus. From Monroe W. Strickberger, Genetics, 2nd ed. (New York: Macmillan, 1976), p. 680. Used by permission.

of the *lac*-operon is found in Figure 12. The regulator gene appears first and is followed by the promoter sequence. In this instance, the promoter also serves as the binding site for a protein called CRP protein. This is a positive regulator because RNA polymerase will not attach without it. Next follows the operator sequence, which is followed by the structural genes. The operon is concluded by the terminator sequence that halts transcription.

Initially, it was thought that all the genes were arranged on the DNA molecule without interruption. The sequence of nucleotides just flowed from one gene to the next. The discovery of regulatory genes, operators, and promoters soon began to change that concept. Later the discovery of repetitive DNA contributed to the changing concept of the structure of DNA in organisms. Repetitive DNA is a segment of DNA that is not transcribed. The repetitive sequence might consist of only one or two nucleotides repeated hundreds of times, or even more. Other sequences consist of a unit of several nucleotides repeated hundreds or even thousands of times. Some suggest that these nontranscribed repetitive sequences may have a regulatory role.

The thought that genes are not lined up sequentially was perplexing, but recent discoveries dwarf it by comparison. It has now been learned that in eukaryotic organisms, the gene itself is interrupted by noncoding nucleotide sequences. After transcription of a long portion of DNA into mRNA, regions of this RNA are spliced out and the remaining portions are fused together to form the mRNA, which is eventually translated into protein. The portions that are spliced out are called introns; the remaining translating regions have been named exons. Vertebrate genes may contain eight, fifteen, or even fifty exons, which are very short coding regions compared to the introns, which may be hundreds or thousands of base pairs in length. Not only, then, are there portions of DNA that are not transcribed, but much of what is transcribed is not translated. Just how the cell knows what to keep and what to throw away is not known. These discoveries were spawned by the relatively new recombinant DNA techniques. This is sure to be an intense area of research in the future. Our discussion so far has centered around the microstructure and function of the hereditary material, DNA. Now we turn our attention to the chromosomes, the cellular bundles for DNA.

The Makeup of DNA

Most of what has been described up to this point is unobservable, even to the most powerful microscope. The chromosomes were observable long before the full nature and role of DNA was understood. Eukaryotic cells package their DNA in the chromosomes inside the nucleus. In prokaryotes, which lack a nucleus, there may be only a single chromosome.

As far as is known, chromosomes are composed of DNA, some RNA, a class of proteins called histones, and another group of heterogeneous proteins simply referred to as nonhistones. This complex is usually referred to as chromatin. It now appears that each chromosome contains a single double-stranded molecule of DNA. Electron micrographs of eukaryotic chromatin show a "beads-on-a-string" substructure. Each bead is a core particle that is an octamer of histones with about a 140-base-pair segment of DNA wrapped around it. Between each core particle is a 15-100-base-pair segment of DNA called a "spacer" or "linker." The core particle plus the linker DNA is called the nucleosome. Further coiling of the solenoid structure produces the densely packed metaphase chromosome (Figure 13).

Chromosomes

During a cell's growth or synthesis cycle, chromosomes are not discernible structures. The

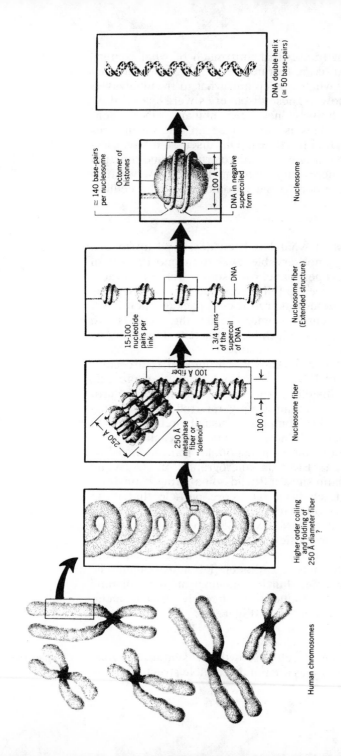

Figure 13. Chromosome Structure: to DNA Level. From Gardner and Snustad, *Principles, inside front cover. Used by permission.*

cell nucleus simply appears as a dark, jumbled mass. During this period, the chromosomes are probably uncoiled, exposing the individual genes for transcription. As the cell prepares to divide, the chromosomes begin to coil and slowly emerge as separate entities. It is during cell division that the number of chromosomes possessed by an organism can be counted. In sexually reproducing organisms, chromosomes occur in pairs. Man, for instance, possesses forty-six chromosomes, twenty-two pairs of autosomes, and one pair of sex chromosomes. Each pair of chromosomes can be identified by their size, position of a constriction called the centromere, and by a banding pattern that emerges upon chemical staining (Figure 14). The nucleus of essentially every cell in the human body contains all forty-six chromosomes. A liver cell, however, functions solely as a liver cell because certain portions of the DNA in the chromosomes are "turned off" by an unrecognized mechanism.

Since each cell contains all the chromosomes, each cell division must equally duplicate and divide the genetic material so that each daughter cell receives the necessary genetic complement. The process of mitosis accomplishes this in eukaryotic organisms. Let's suppose that an organism possesses four chromosomes: one pair of short chromosomes and one pair of long chromosomes (Figure 15). As the cell prepares to divide, each chromosome is duplicated. The sister chromosomes (chromatids) remain joined at the centromere. The sister chromatids separate as the cell reaches the final stages of division. Each daughter cell then winds up with an exact duplicate complement of chromosomes as the parent cell has. It appears that the sole function of mitosis is to insure efficient and accurate transmission of genetic material from parent to daughter cells.

Organisms that reproduce this way are asexual. Sexually reproducing organisms are those in

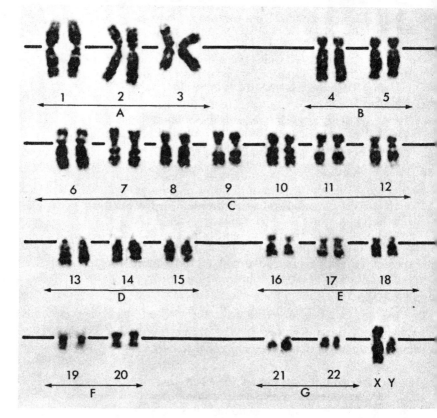

Figure 14. Human Karyotype G-banding. Source: Paris Conference (1971): Stadardization in Human Cytogenetics *(White Plains, N.Y.: National Foundation March of Dimes, 1973), p. 14. Used by permission.*

which there is a union of two gametes to form a zygote. Since the zygote must contain the same number of chromosomes as the parents (diploid number), the gametes, therefore, contain only half (haploid number) of the chromosomes. In humans, for example, the diploid number is forty-six, and the haploid number is twenty-three. Each gamete or sex cell contains one chromosome from each of the twenty-two pairs of autosomes and one sex chromosome. The process of halving the chromosome number in gametes is known as meiosis.

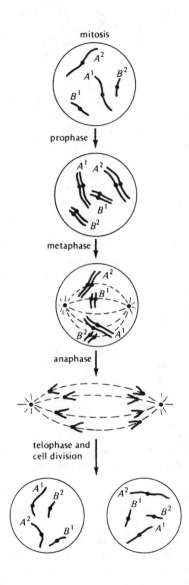

Figure 15. Mitosis. From Strickberger, Genetics, *p. 23.*

Meiosis takes place in two divisions. Whereas mitosis yields two diploid daughter cells, meiosis yields four haploid gametes. The first meiotic division is reductional. During this division, the chromosomes duplicate as in mitosis, but the homologues also pair up—a feature unique to meiosis. At the division, the homologous pairs separate again. The resulting daughter cells contain (in humans) twenty-three duplicated chromosomes. The second equational division simply splits the twenty-three duplicated chromosomes. The end result is four gametes with twenty-three chromosomes (Figure 16).

There are two important events in meiosis that, by themselves, create a staggering amount of variation between organisms of the same species. The first is the random assortment of paternal and maternal chromosomes. Each of us received a set of twenty-three chromosomes from our father (paternal) and another set from our mother (maternal). During the first meiotic division, when the homologues split up, the maternal homologue is separating from the paternal homologue. The separation occurs randomly for each pair. Since each pair can split one of two ways, and there are twenty-three pairs, there are 2^{23} possible gametes, or over 8 million (8,388,608) possible gametes simply by random assortment of maternal and paternal sets of chromosomes at meiosis. That means that a married couple has the possibility of producing over 70 trillion different children by this process alone. (8,388,608 x 8,388,608). No wonder, apart from identical twins, no two siblings look exactly alike. This is the process of random assortment.

In addition, during the first meiotic division, when the homologues are paired, the chromatids may actually exchange pieces with each other (Figure 17). Crossing over adds further to the shuffling and, hence, to the possible number of gametes that can be produced. Therefore, not only are the paternal and maternal chromosomes

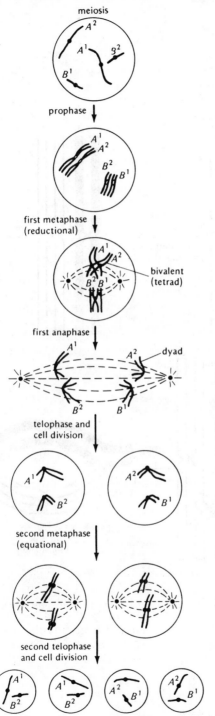

Figure 16. Meiosis.
From Strickberger
Genetics, p. 23.

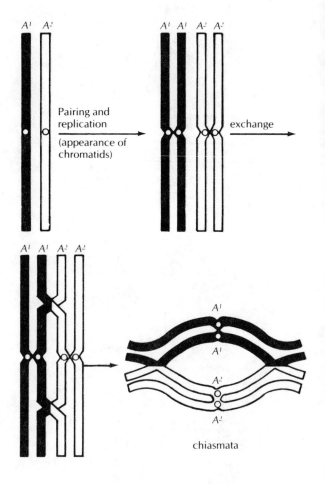

Figure 17. Chromosome Exchange. After Strickberger, Genetics, p. 20.

separated but they also exchange pieces with each other. Recombination, even more than random assortment is the process responsible for the uniqueness of different individuals of the same species. New combinations of genes or chromosomes are constantly formed. The odds that the end-products of any two series of meiotic division will be the same are practically nil. It has been estimated that each human individual is capable of producing more different germ cells than there are atoms in the universe (estimated at 10^{80}).

At the moment, it would seem as if the system were perfect. The mechanism of DNA replication assures accuracy, as does transcription and translation. Random assortment and recombination provide a mechanism for maintaining variability and supplying new combinations. But mistakes do occur. These mistakes are known as mutations.

Mutations

Most mutations take place during DNA replication. There is a change, addition, or deletion of a single nucleotide. These point mutations, though seemingly insignificant, can have profound effects. By changing or substituting one nucleotide for another, the nature of the triplet is altered. This alteration could result in a different amino acid being placed in the protein chain during protein synthesis. For example, take AUA, which codes for the amino acid isoleucine. It should be noted that we are using the conventional mRNA designations for codons, though DNA is normally the site of mutagenesis. If adenine substitutes for uracil, the resulting mRNA codon translates into lysine, a different amino acid. This could change the entire structure and function of the resultant protein molecule. But because of the redundancy of the genetic code, this may not always be the case. If AUA is changed to AUU, isoleucine is still inserted. Or the substitution of lysine for isoleucine may not affect the protein in any significant manner.

A nucleotide substitution that displaces one amino acid for another is called a missense mutation. If a stop codon results, a mutation has occurred, because the stop codon terminates all messages. This is a nonsense mutation. If a nucleotide is added or deleted, this creates a more significant problem. The reading frame shifts for the entire remainder of the molecule. Consider this sequence of nucleotides: ACU ACA UGG CAU GUC—which codes for threonine (ACU), threonine (ACA), tryptophan (UGG), histidine (CAU), and valine (GUC). If a uracil (U) is added between cytosine and uracil of the first codon, the change is dramatic: ACU UAC AUG GCA UUU C. The first amino acid remains threonine (ACU), the second changes to tyrosine (UAC), the third changes to the methionine (start) codon (AUG), the fourth to alanine (GCA), the fifth to cysteine (UGU). This type of mutation results in a rash of missense and nonsense mutations, rendering the resulting protein useless.

When point mutations occur in somatic cells, only the individual organism is affected. If the mutation occurs in the germ cell, provided the function of the sperm or egg is not sabotaged, the mutation may pass on to the next generation, though not expressed in the individual in which it arose. Point mutations are generally detected by an alteration in the function of an enzyme or other protein indicating a mutational change. If such a change is not apparent, various biochemical assays and amino acid or nucleotide sequencing techniques are necessary to detect the mutation.

Chromosomal mutations present different problems. They sometimes result in large-scale malformations that are easily identified, especially those that leave an individual with fewer or additional whole chromosomes. When this is not the case, chromosomal mutations are sometimes detectable microscopically, using various staining techniques applied during cell division when the chromosomes are large and distinct.

A segment of a chromosome may be duplicated (duplication). One or more chromosomes may be fused together (fusion), thus reducing the number of chromosomes. A segment of a chromosome may be deleted (deletion), inserted (insertion), or inverted (inversion). A segment of a chromosome may be transferred to an entirely different chromosome (translocation) or to a different position on the same chromosome (transposition). Though some chromosomal mutations are easily detectable, the piece of chromosomal material that is actually rearranged may be so small as to render it impossible to distinguish from a point mutation.

As can be readily seen, mutation is a very complex phenomenon. Its many faces appear to rise from numerous directions. But it must be remembered that mutations rarely occur. They are the exception rather than the rule. Considering a host of both eukaryotic and prokaryotic organisms, the chances of a single gamete containing a new mutation for a particular gene range from 1/2000 to 1/1,000,000,000. In the drosophila fruit fly, mutations for a single gene average around 1 per 100,000 gametes. In general, mutation rates fall in this range plus or minus one order of magnitude. In approaching the subject of genetic variation, our next topic, it is necessary to understand mutation.

To understand the terminology of genetic variation, we must return for a moment to chromosomal theory. Aside from random assortment of cell division, another consequence of chromosomes occurring in pairs in sexually reproducing organisms is that genes also occur in pairs, one gene each on identical locations of the homologous pair. This point on the chromosome is referred to as the locus (pl. loci). With

The Measure and Nature of Genetic Variation

each individual carrying two genes for the same trait, this raises the possibility of having these two genes being slightly different. While they still code for the same protein, a slight difference in the nucleotide sequence will produce a protein that still performs the same basic function yet is slightly different structurally. These alternate expressions of a gene are termed alleles. When members of a gene pair in an organism are identical, the organism is considered homozygous for that gene pair. If the alleles are different, the organism is considered heterozygous. Examples are easily seen in Mendel's yellow and green peas, brown versus blue eye color in humans, and brown versus black coat color in guinea pigs. In each of these examples, there are two alleles. Any one guinea pig, for instance, would be either homozygous for black, heterozygous for black and brown, or homozygous for brown.

Genes with more than one allele are said to be variable; hence the term *genic variation* or in a broader sense, *genetic variation*. Genetic variation in natural populations has been the subject of intense investigations ever since the confirmation of Mendel's work at the turn of the century. It is now clear that genetic variation is much more pervasive than previously imagined in terms of the proportion of genes with more than one allele, the number of alleles per individual locus, the number of genes affecting a single trait, the number of heterozygous pairs per individual, and the number of individuals heterozygous at a particular locus.

Prior to the 1950s and 60s, the "classical" theory held that variation was very rare. The gene pool of a population was thought to consist of a predominant wild-type allele at each locus, with only a few variant alleles. Every individual was homozygous at practically every locus for the wild type. In fact, "too much" variation was thought to create a "genetic load." Classical theorists drew this conclusion primarily from the extensive studies of the fruit fly *Drosophila* and its easily observable, yet rare morphological mutations. As new techniques arose to study the fine structure of chromosomes, the mystery of DNA and its code gradually unraveled. Furthermore, when advanced methods of testing for differences in protein emerged, the estimated levels of variation in natural populations continually rose.

The most influential technique, gel electrophoresis (Figure 18), was first applied to population genetics in 1966, by Lewontin and Hubby.[3] In this process, a small tissue sample is taken from an organism and homogenized to release the proteins. Next, the tissue samples are placed on a gel of starch, agar, or polyacrylamide. The gel is then subjected to an electrical current for a few hours. Every protein will migrate in this electrical field at a rate that is dependent on its net electrical charge and molecular size. The gel is then treated with a chemical solution that causes the location of the protein (enzyme) to become visible as a colored band or series of bands. Enzymes produced by the differing alleles may have different molecular structures and electrical charges because of differences in amino acid composition; consequently, the alleles will travel greater or lesser distances on the gel. Thus one can detect genetic variations at the protein level by measuring the number and position of these bands.

The ability to detect genetic variation at the protein level opened a whole new field of investigation. In the ensuing years and up to the present, gel electrophoresis has been applied to other invertebrates, plants, and representatives from all five classes of vertebrates at dozens of gene loci. According to the classical model, individuals would be expected to be heterozygous at less than 1 percent of all loci. This percentage is based primarily on the rates of mutation. Now we know that in most natural populations an individual can be heterozygous

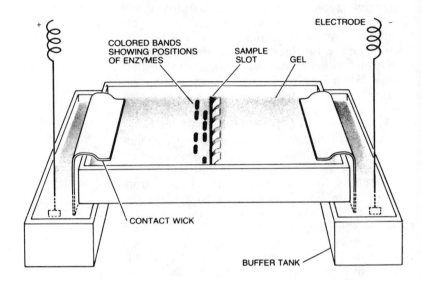

Figure 18. Laboratory Technique: Gel Electrophoresis. Gel electroporesis is a method for estimating the genetic variation of natural populations by examining the variant proteins manufactured by different individuals. First a tissue sample from each of the organisms to be surveyed is homogenized to release the proteins in the tissue; the proteins are placed on a gel of starch, agar or polyacrylamide. The gel with the tissue samples is then subjected to an electric current, usually for a few hours. Each protein in the sample migrates through the gel in a direction and at a rate that depend on its net electric charge and molecular size. After the run is over the gel is treated with a chemical solution containing a substrate that is specific for the enzyme under study and a salt. The enzyme catalyzes the conversion of the substrate into a product, which then couples with the salt to give colored bands at the points to which the enzyme had migrated. Because enzymes that are specified by different alleles may have different molecular structures and charges (and hence different mobilities in an electric field) the genetic makeup at the gene locus coding for a given enzyme can be established for each individual from the number and position of the electrophoretic bands. Source: "Mechanisms of Evolution," Scientific American *(November 1978), p. 61. Used by permission.*

for 5-15 percent of all loci, one to two orders of magnitude greater. This caused a rethinking of the genetic structure of natural populations.

Natural populations are now believed to store a large proportion of genetic variation. The variation is seen as a buffer against minor environmental fluctuations. This "balance" model views variation as a "hedge against inflation." Alternate expressions of genes may help an organism to survive expected (seasonal) temporary environmental fluctuations, as well as unexpected, permanent changes.

The success of gel electrophoresis to provide new insights motivated other researchers to devise new methods of analyzing genetic variation. This led, first, to determining amino acid sequences of proteins and, second, to the determination of nucleotide sequence of the actual gene. Since electrophoresis depends on changes in molecular structure and electrical charge, not all amino acid substitutions are detectable. Therefore, sequencing the protein will reveal all differences in amino acid sequence between two alleles. Amino acid sequencing provides a means of detecting all variations at the protein level, but nucleic acid sequencing provides a true measure of all genetic variation. Since the triplet code is redundant, a nucleotide substitution will not always result in an amino acid substitution. Therefore, to measure the full extent of genetic variation, the gene itself must be sequenced.

Nucleotide sequencing, however, has proved to be more complicated than anticipated with the discovery of introns, sections of meaningless DNA, found in the midst of the gene. As mentioned earlier, the introns are taken out of the mRNA before it is translated into protein. Through nucleotide sequencing, it has been found that introns contain a greater degree of variation than do exons. The function of introns is still open to debate, and the debate is intensified by the realization that mitochondrial DNA contains no introns and half the genes do not even possess a stop codon.

Though amino acid sequencing and nucleotide sequencing may be the waves of the future, they are still somewhat time-consuming and expensive. Therefore, at present, their use is limited to comparisons of the genetic differences between species and not to comparisons of interspecific and intraspecific variations within populations of a single species. Electrophoresis remains the most effective tool in this study.

Now that a basic framework in modern genetics has been laid, we would like to turn our attention to the major theme of this book: the nature of biological change.

chapter four

Neo-Darwinism

Neo-Darwinism has been the major theory of biological change for over half a century. The basic tenets of Neo-Darwinism will be outlined.

The origin of adaptation and the nature of genetic variation is currently interpreted under the all-encompassing concept of evolution. In fact, Theodosius Dobzhansky has said that "nothing in biology makes sense except in the light of evolution."[4] Though the majority of biologists would agree with that statement, there is considerable disagreement over just what evolutionary process is responsible for the major proportion of evolutionary change. At present, there are two primary camps. The Neo-Darwinians believe that evolutionary change is slow, gradual, and evenly distributed over time. Barring dramatic environmental alterations, few

65

events are believed capable of altering this slow steady pace. On the other side stand the punctuationalists. The punctuationalist theorizes that evolution takes place in relatively sudden spurts followed by long periods of stasis. A small population may branch off, rapidly speciate, and quickly adapt to a slightly different environment than the larger ancestral population. This will be followed by a long period of stability with little if any evolutionary change until the next speciation event takes place.

Since there is always a positive and a negative aspect to every controversy, the ensuing chapters will attempt to present both sides of each alternative, one chapter to explain the skeleton of each theory, along with its baggage of logic and evidence, and a second chapter to dissect what the critics have labeled fallacious, exaggerated, or misunderstood. Though no one can claim to be totally objective in such a discussion, our hope is that both sides will observe that they are fairly and adequately (though necessarily incompletely) represented. Our attention will be initially turned to the older of the two, Neo-Darwinism.

The Origins of Neo-Darwinism

Neo-Darwinism is also known as the modern synthesis or the synthetic theory. Its beginnings are found in Dobzhansky's *Genetics and the Origin of Species*[5] written in 1937. This was followed closely with major works by Huxley,[6] Mayr,[7] Simpson,[8] and Stebbins.[9] The synthetic theory drew together information from population studies, Mendelian genetics, mutation theory, and chromosome theory. Essentially, the synthetic theory views the origin of novel adaptive structures through the interaction of mutation and natural selection. Genetic variation in natural populations serves as the raw material on which natural selection operates. Mutations serve as the ultimate source of all genetic variation. To the Neo-Darwinist, evolutionary change results from the interaction of genetic variation (originating by mutation) and natural selection.

Mutations and Evolution

With our modern understanding of genetics, it becomes clear that any evolutionary system must depend on mutations as the ultimate source of all genetic variation and, hence, all biological novelty. Mutations are mistakes, errors in the precise machinery of DNA replication. Combine this with the rarity and randomness of mutations, and one has a major reason why Neo-Darwinists perceive evolutionary change as being gradual and slow. Since any specific mutation is rare, and most are deleterious, a mutation that somehow enhances survival is admittedly highly unlikely, though not impossible.

In fact, the rarity of a particular mutation in an individual contrasts with the extent of mutations actually present in large populations. For instance, in the human population, the average mutation rate is estimated at 10^{-5} per gene per generation. McKusick suggests that humans possess 100,000 pairs of genes or 2×10^5 genes. Multiplying the figures generates the probability that on the average, each human being contains two newly generated mutations. Multiplying this by the current world population of 4×10^9, one realizes that the human population currently contains 8×10^9 newly arisen mutations. An average of 80,000 mutations arise at each locus per generation ($4 \times 10^9 \times 2$ genes $\times 10^{-5}$).[10] Consequently, the mutation is rare in the context of a single gene in an individual, but when populations are considered, mutations appear to be a rather pervasive phenomenon. Certainly among all these mistakes, the Neo-Darwinist argues, something beneficial will be presented as well.[11]

The randomness of mutations is another phrase that must be properly understood. F. Y. Ayala explains:

The forces that give rise to gene mutations operate at random in the sense that genetic mutations occur without reference to their future adaptiveness in the environment. In other words, a mutant individual is no more likely to appear in an environment in which it would be favored than in one in which it would be selected against. If a favored mutation does appear, it can be viewed as exhibiting a "preadaptation" to that particular environment: it did not arise as an adaptive response, but rather proved to be adaptive after it appeared.[12]

Randomness demonstrates itself as well in the context of predictability. It is impossible to predict when a particular gene will mutate; it is possible to predict only that there is a certain probability that a mutation event will occur. Randomness does not mean, however, that one mutation is as likely as any other. Nucleotide substitutions, additions, and deletions may have different probabilities of occurrence. Therefore it should be clear that when the evolutionist speaks of the randomness of mutations, it does not mean that mutations can occur anywhere, anytime. There is a degree of regularity and predictability. Randomness most often means that mutations do not anticipate or respond to an environmental condition but instead are oblivious to the survival needs of an organism.

Another element of the gradual nature characteristic of Neo-Darwinian theory involves the realization that mutations occur in organisms that are already adapted to their environment. Almost any change in an already adapted, well-functioning system will be deleterious. Not only does this hint at the rarity of beneficial mutations but also, according to Neo-Darwinians, these few rare beneficial mutations will bring about only minor adjustments in the organism and its life history pattern. Any large-scale, rapid alteration to the organism will not only be deleterious but most likely lethal. The analogy has often been made to that of a finely made clock. The removal of a gear will cause the clock to cease functioning. It would be more accurate to say that major evolutionary change occurs by the gradual accumulation of minor point mutations analogous to the tightening of a screw in the delicate mechanism of nature.

Two qualifications must be made at this point. First, not all biological change is dependent on new mutations. As will be discussed later, there already exists a great storehouse of genetic variation in natural populations. By simply reshuffling this variation, principally through recombination and migration, quite dramatic changes result. In regard to the fruit fly *Drosophila*, "There appears to be no character—morphogenetic, behavioral, physiological, or cytological—that cannot be selected in *Drosophila*."[13]

But this sort of reshuffling is not likely to result in major evolutionary change. It simply reorganizes what already exists. Even with new mutations, it will not likely lead to new original metabolic functions and new protein molecules. This brings us to the second qualifier. Since the development of whole new nucleotide sequences is highly unlikely, it is suggested that new genetic information gets its start from a preexisting gene that is duplicated. Newly duplicated genes can then be modified to slightly different genes with slightly different functions through the build-up of mutations that are unique to it.[14] Duplication is not limited to a single gene; one or more chromosomes or even the whole genome can be duplicated, opening up vast amounts of genetic information for modification. The most well-known example of gene duplication involves myoglobin and the four hemoglobins. Myoglobin is said to be the ancestral protein with α, β, γ, and δ polypeptide chains resulting from gene duplication and subsequent evolution. Whole phylogenetic trees have been worked out for these five proteins.[15]

Mutations, genetic variation, and recombination by themselves will not generate major evolutionary change. These phenomena will not produce a bird where there was once a reptile, no matter how much time is given. There must be a process by which new adaptive mutations can be preserved and spread to the entire population. That process is called natural selection. When Darwin proposed his theory of natural selection, he made three main observations. First, variation exists in all natural populations. Second, most species can reproduce beyond the capacity of their resources. Third, resources are limited. Therefore, individuals with the most advantageous variations will survive. These observations still hold true today. In Neo-Darwinism, natural selection is defined in terms of differential reproduction, rather than mere survival of the individual alone. Those variations that allow individuals to reproduce in proportionately greater numbers will in time come to dominate the species. Darwinian fitness is no longer defined in terms of superior structural, functional, or behavioral adaptation. Nor is it defined in terms of the number of offspring produced. Rather, it is understood solely in terms of a greater proportion of a genotype in succeeding generations.

Natural selection, as it is understood today, is a complex phenomenon. There are many different varieties of natural selection, each with its own role of either preventing or promoting biological change and maintaining or eliminating genetic variability. Current theories of natural selection present three basic categories: stabilizing selection, balancing selection, and directional selection. *Stabilizing selection,* sometimes called normalizing selection, represents the conservative aspect of selection. As the name implies, stabilizing selection describes the process by which a population maintains its adaptiveness by eliminating deleterious variants:

Normalizing natural selection is preeminently a conservative force. It purges the gene pool of a population of deleterious genetic variants and thereby tends to keep the species constant.[16]

Balancing selection encompasses two different types of selection, all of which involve the maintenance of genetic polymorphisms in natural populations. Polymorphisms exist when two or more genetic variants occur in a population at frequencies higher than can be accounted for by random processes. Human eye color exhibits polymorphism with the presence of blue, brown, green, grey, and hazel colors. Strict balancing selection occurs when the heterozygote exhibits a higher fitness. The classical example in man is sickle-cell anemia. In certain African and Asian populations, there are two alleles for hemoglobin. The S allele codes for normal hemoglobin; the s allele codes for the mutant, sickle-cell hemoglobin. The anemia is due to homozygosis of the s allele. The s/s homozygote suffers chronic health problems and higher mortality. The S/s heterozygote not only survives but is also resistant to certain forms of malaria that the normal homozygote S/S is not. Therefore, the heterozygote has a distinct advantage over the two homozygotes. In areas where malaria is present, the heterozygote leaves more survivors than either competing homozygote, thus ensuring that both alleles remain in the population.

The second form of balancing selection is diversifying (also called disruptive) selection. When a population inhabits an environment that is heterogeneous, to sustain a high level of adaptiveness requires some degree of genetic variation be maintained in the population as a whole. This is diversifying selection. Although it has been difficult to demonstrate diversifying selection in nature (Darwin's finches are sometimes offered as a possibility), laboratory studies with *Drosophila* indicate populations placed in heterogeneous environments support more genetic polymorphism than those in homogeneous environments.[17]

The third major type of natural selection is *directional selection*. Directional selection describes the events that lead to biological change. Normalizing and balancing selection are generally conceived as conservative forces. Directional selection describes the creative role of natural selection. Either of two conditions will lead to directional selection. First, if a genetically variable species is exposed to a new environment, some genetic frequencies may begin to change in response to the environmental shift. Second, if the environment remains constant, but new favorable mutant genes occur, this may give the population a higher level of adaptiveness. In either case, the genetic structure of the population changes over time and moves to a new level of adaptiveness. This is the natural selection that Darwin had in mind.

In summary, in regard to natural selection, it should be pointed out that natural selection is not an active process, but passive. There is no selector. These different types of selection are simply categories of descriptive events. Natural selection does not select anything; it simply happens. In that sense, it is probably misleading to assign natural selection a truly creative role because it can only deal with the genetic variation already present. Though this creative role is frequently mentioned, it must be understood in a figurative sense. It is creative only in the sense that it is a process that may lead to new gene combinations. Also, it is frequently stated that natural selection brings a population to the *most* adapted state. Natural selection maximizes or optimizes one trait or another. This also is misleading. The optimal solution might not be genetically available. Selection merely arrives at or preserves the nearest adaptive state and not necessarily the best.[18]

So far, we have said little concerning the role of speciation in the Neo-Darwinian model. This is because speciation has played a rather secondary role. This is not to say that it is not important. If speciation did not occur, then no matter what evolutionary mechanism one arrived at, there would still be only one or a few species on the planet. Speciation provides diversity, much as gene duplication does. Once there are two species, two separate gene pools that are no longer interbreeding, they can then travel completely separate evolutionary paths, affected by different environments and different mutations. Since punctuated equilibrium maintains that speciation events play a central role in bringing about major evolutionary change, speciation phenomena will therefore be discussed more fully in that context. In the Neo-Darwinian model, suffice it to say that speciation provides the opportunity for diversity and is itself the product of gradually accumulating minor changes. It is able to play only an indirect role in bringing about major evolutionary change.

**Mutation and
Natural Selection**

In order for Neo-Darwinian evolution to be considered a viable mechanism, it must be able to supply examples of biological change taking place by the action of mutation and natural selection in both natural and experimental populations. We will discuss five examples that are commonly cited: bacterial antibiotic resistance, insect pesticide resistance, industrial melanism, sickle-cell anemia, and increased fitness in irradiated populations of *Drosophila*.

It is now a well-known phenomenon that pathogenic and other bacterial populations can become increasingly resistant to antibiotics, such as penicillin or streptomycin. The earliest work in this area began in the 1940s.[19] Whenever a culture of bacteria was exposed to an antibiotic, there would always be found a small number of cells that were resistant to the antibiotic.

These resistant cells were originally thought to be the result of favorable mutations only, but it is now known that many are the result of gene transfer between cells. By increasing the dosage of antibiotic step by step, one could eventually select for bacteria that were resistant to even the greatest concentration of the antibiotic. This stepwise pattern led many researchers to conclude that there were a number of genes involved and that the stepwise resistance pattern resulted from additional mutations in the other genes involved.

An analogous situation is the resistance of insects to insecticides. Hundreds of different species of insects have developed some degree of resistance to insecticides.[20] The pathways of resistance to different chemicals appear to be different from each other, and a stepwise progression of increasing resistance can be obtained. Resistance has occurred independently in insect populations in different parts of the world for the same chemical. This indicates that the resistance is arising independently and is not being spread from one population to another. However, resistance may be due to the large effect of a single gene or to the combined effect of many genes with individually minor effects.

The classic example of natural selection that is found in all biology texts from junior high to the graduate level in college is industrial melanism in moths, studied mostly in England.[21] In the industrial areas of Britain, over one hundred species of moths have increased the proportion of the darkly pigmented variety and replaced the light variety in some areas. The best known, however, is *Biston betularia*, the peppered moth. Before 1848, almost all specimens of the peppered moth were of the greyish, lightly speckled variety. The dark, almost black melanic form increased in frequency until 1895, when it comprised 98 percent of the population.

Prior to the industrial revolution, the trees of Britain's forests were covered with light-colored lichens, on which the peppered variety was nearly invisible. As air pollution levels rose, the lichens began to disappear, leaving the trees very dark. The melanic form of the moth, which had been seen only rarely before, began to increase in frequency. Previously, the melanic form was very conspicuous as it rested on the lichen-covered bark and was therefore susceptible to heavy predation by birds, whereas the light, peppered variety was concealed. As pollution killed off the lichens, the situation reversed. The melanic form was now better camouflaged and escaped predation at a greater frequency. In most species, the change from light to dark was suspected to be the result of a single gene mutation that was dominant. A further interesting note is that since the pollution levels have started to drop, the lichens are making a comeback, and so is the light, peppered variety.

Although these three examples are all the result of human intervention, they are all clear examples of directional selection. When a rare allele is exposed to a new environment, it offers a distinct advantage, whereas in the previous environment, it would have been deleterious or neutral at best. With continued selection pressure from the environment, the new genetic variant soon dominates the entire population. The evolutionary sequence is complete.

The case of sickle-cell anemia presents a much different situation. As an example of heterosis, or balancing selection, the genetic mutant will not be able to become fixed throughout the entire population in any environment. Since the homozygote for the sickle-cell gene is unhealthful and sometimes lethal, independent of the environment, an environmental change will never allow this allele to predominate. The mutation that results in a single amino acid change in a hemoglobin does offer an advantage, but only in the heterozygous condition and only in the presence of malaria. So, although the sickle-cell allele can never eliminate the normal allele in the population, it is an example of a mutation providing a beneficial effect in a particular environment.

The last example involves the influence of artificially caused mutations on the fitness of a population. There is experimental evidence that points to a positive correlation between genetic variability and the rate of evolutionary change.[22] Hybrid populations of fruit flies were able to adjust to a stable laboratory environment much quicker than the two single strains used to form the hybrid. The reason was the higher level of genetic variation present in the hybrid population than in that of either of the parent strains.

Another means besides hybridization for increasing genetic variability is that of inducing mutations. Since all new genetic variation ultimately arises by mutation, perhaps favorable genetic mutants can be produced artificially. Francisco Ayala, a population geneticist at the University of California (Davis), performed such experiments with *Drosophila birchii* in the late 1960s.[23] The males of a population were irradiated in each of the first three generations of the laboratory study. The rate of adaptedness or increase in fitness was measured by the increase in population numbers in an environment of constant space, temperature, and food availability. Three pairs of experimental populations were established from natural populations from Australia, New Zealand, and New Britain. One population from each pair was not irradiated, and thus served as a control. Following the initial population crash following exposure to x-rays, the experimental populations eventually caught up to and surpassed the control populations in all three pairs. Although all six populations increased in fitness as adjustment to the conditions took place, the irradiated populations increased faster. Ayala concluded that along with the deleterious mutants, which were quickly eliminated, a few favorable mutations were produced in the surviving flies. This increased the genetic variation in these populations, allowing them to adapt more quickly to a new environment.

These classic examples are used to show that the phenomena of mutation and natural selection can bring about changes that enhance survival in a population. However, it is easy to misinterpret these examples as demonstrating evolution through the mutation of only a single gene. The Neo-Darwinist rightly emphasizes that genes not only act on their own but also interact with other genes. Dobzhansky states:

> Darwinian fitness is a property not of a gene, but of a genotype and of the phenotypes conditioned by this genotype. Selection favors, or discriminates against, genotypes, that is, gene patterns.[24]

The expression of a gene can, therefore, be a complex phenomenon. Ultimately, of course, it is the phenotype that is selected. But it is the genotype that is tracked through time. It should be concluded, then, that the origin of new adaptations should not be expected to arise by the mutation of a single gene: there will be other factors involved. For instance, the melanic form of *Biston betularia* was crossed with a light-colored, related American species. The result was not a distinct light and dark form, as in Britain, but a shaded series from light to dark. This indicated a more complex genetic system than was first thought.

Genic Variation and Neo-Darwinism

Another mistaken conclusion is that the mutation process is necessary for any evolutionary change to take place. Neo-Darwinists believe that although mutations are the ultimate source of variability, much biological change is possible without mutations, as the late Theodosius Dobzhansky clearly stated:

> That selection can work only with raw materials arisen ultimately by mutation is manifestly true. But it is also true that populations, particularly

those of diploid outbreeding species, have stored in them a profusion of genetic variability. A temporary suppression of the mutation process, even if it could be brought about, would have no immediate effect on evolutionary plasticity. Rapidly evolving groups need not have high mutation rates, nor should evolutionary stasis be taken as evidence of insufficient mutability.[25]

The extent of genetic variation in natural populations plays an important part in the Neo-Darwinian model. The evidence for the pervasiveness of genetic variation comes mainly from three sources: the results of inbreeding, the success of artificial selection, and recent measurements of molecular variation by gel electrophoresis and amino acid and nucleotide sequencing.

It is well known that inbred populations exhibit certain traits that are not found in normal outbred populations. This is because breeding among closely related individuals increases the probability of matching up recessive alleles. In normal outbreeding populations, these recessive alleles, often deleterious, are masked by normal dominant alleles. Inbreeding in laboratory populations is consistently used to uncover unusual recessive traits that a population carries in its gene pool but rarely expresses. This is the first clue that not all the genes in a species are of a single wild type.

The second clue comes from artificial selection experiments. Darwin used artificial selection both as evidence for the presence of variation in species and as a model of natural selection. Man has practiced artificial selection for desirable traits in domestic plants and animals for centuries. The average egg production of a white leghorn flock was increased from 125.6 eggs per hen per year in 1933 to 249.6 eggs per year by 1965, a 100 percent increase in thirty-two years.[26] Selection for over fifty generations in corn brought about large changes in oil and protein content. From a starting point of

10.9 percent, protein content could be brought as low as 4.9 percent and as high as 19.4 percent. Oil content starting at 4.7 percent was raised to 15.4 percent and lowered to 1.0 percent Even after these fifty generations, it was evident that there was still variation present. A limit or plateau had not yet been reached.[27] *Drosophila melanogaster* had been subjected to artificial selection for over fifty-one different traits, both morphological and behavioral. It has been stated repeatedly that almost any trait is changeable if selected for rigorously and over many generations.

In recent years even more convincing evidence of the extent of genetic variation has surfaced at the molecular or protein level. The utilization of gel electrophoresis has made possible the measure of variation that was previously undetectable. It must be remembered, though, that electrophoresis does not detect all amino acid substitutions. It detects only those that result in a change in electrical charge or a changed tertiary or quaternary structure. Not all amino acid substitutions will produce such changes. Consequently, even though electrophoresis has increased the estimates of variation in natural populations, it is by no means complete. King and Wilson estimated that only 27 percent of amino acid substitutions are detectable by electrophoresis, though some have recently speculated that the percentage may be much higher.[28]

As would be expected, the amount of variation differs from protein to protein and species to species. Differences in the levels of variation between proteins appear to be correlated with metabolic function. Enzymes involved in the glucose metabolizing pathway (Group I) tend to be less variable than other less specific enzymes (Group II). Nonenzymatic proteins (Group III) show even less variability. "Mean individual heterozygosity (H) is in all groups of organisms roughly twice as large for Group II enzymes as for those in Group I."[29] (H is an estimate of how many loci in an individual are likely to be heterozygous. This figure ranges from 0 to 25 percent but is usually between 5 and 15 percent.)

Levels of heterozygosity (H) and polymorphism (P) differ also between taxonomic groups. (P is an estimate of the number of loci for which a population is polymorphic.) Insects and plants exhibit the highest levels of variability, mammals and birds, the least. It is difficult at this time to draw firm conclusions from this data. The data in Figure 20 is from less than 140 species and no more than 40 gene loci per species. The major point is simply that there is a great deal of genetic variability in most natural populations. As mentioned in chapter 2, the average human is estimated to have 6.7 percent heterozygous loci. That means that if humans have 100,000 pairs of genes, 6,700 loci are heterozygous!

Other, more expensive, techniques are indicating, as expected, that gel electrophoresis does not indicate the whole picture. Amino acid and nucleotide sequencing are revealing variation that electrophoresis is unable to detect. It is hoped that these techniques can lead to a greater understanding of evolutionary relationships and genetic plasticity.[30] As new and better techniques are developed to measure genetic variation, such variation will continue to hold the prominent place in the Neo-Darwinian model.

Stochastic Events

One feature that has long been neglected by Neo-Darwinists but is gaining increasing recognition is the stochastic element. Stochastic processes are those events of nature that are more probabilistic than deterministic. These are genetic drift, bottlenecking, and the founder principle. All of these events essentially deal with small populations where random environmental circumstances can have the greatest effect. Genetic drift is a change in gene frequencies due to the effects of random sampling of gametes during reproduction in small populations. Bottlenecking alters gene frequencies when the population experiences a crash phase and the remnant population survives only by chance. This leaves a nonrepresentative sample of the gene pool to carry on. The founder principle provides for random change when a few individuals found a new population isolated from the parent population, as, for example, on an island. Similar to the bottlenecking population, this founder population will most likely contain a nonrepresentative sample of the gene pool simply through the "luck of the draw."

This concludes our cursory description of the Neo-Darwinian model. It appears to present a very persuasive case. However, Darwinism and the more sophisticated Neo-Darwinism have had a stormy history. Not a decade has gone by that it has not been beleaguered by articulate critics. Some of these criticisms have stood the test of time; others have not. In the next chapter we will review what continue to be some of the major problems and criticisms of the Neo-Darwinian model, as well as some new questions that have been asked only in the past decade.

chapter five

Neo-Darwinism Under Attack

Neo-Darwinism is found to contain crucial short-comings in explaining significant biological change.

Some eminent biologists have recently asserted that the problem of evolution is now resolved, except for minor details. This assertion is erroneous. To be sure, there is no reasonable doubt that the living world is a product and outcome of the 3 to 4 billion years of the earth's evolutionary history. However, the causes of evolution and the patterning of the processes that bring it about are far from completely understood.[31]

The uncertainty of evolutionists over the patterns and processes of evolution is as great today as in the first decades of this century. After the 1930s and 40s, when the Modern Synthesis took hold, Neo-Darwinism was defended as if it were dogma. Although it had always had its critics, Neo-Darwinism came under intense attack in the past decade. The 83 complacency of many evolutionists has been shattered. Evolutionary biologists have been forced to reevaluate concepts that have lain unchallenged for years.

Many view this as a sign of health in evolutionary biology. After all, what would scientists do if they had no more questions to ask? The attitude of most evolutionists is similar to that of Dobzhansky et al., quoted above. Evolution itself is not questioned, just its mechanism. It is with this spirit that the majority of comments and quotations in this chapter are to be taken.

Major Areas of Criticism

Our summary of the criticisms of Neo-Darwinism will center on four major areas. The first two are mutation and natural selection. These two processes have been focal points of controversy since the beginnings of the Modern Synthesis. In the third and fourth major areas, we will consider population genetics and paleontology and their role in determining the value of Neo-Darwinism.

Criticism of Mutation

As mentioned in the previous chapter, mutations, ultimately, are necessary for any evolutionary mechanism. There must be some means of bringing about a change in the genetic structure of an organism to produce novelty. Shuffling as in recombination is helpful, but it is not sufficient in itself. One could estimate that beneficial mutations occur only once every 100,000 mutations. A population may have as many as one beneficial mutation occurring in a population per generation. This may not sound like much, but when the gradual evolutionist begins thinking in terms of thousands of generations, the impossible soon becomes merely improbable, the improbable probable, the probable possible, and the possible virtually certain.

Mathematicians, drawn in by the statistical nature of the problem, have denied the feasibility of random minor mutations producing biological novelty and complexity. Using computers, mathematician Marcel Schutzenberger, found that the odds[32] against improving meaningful information by random changes were $10^{-1,000}$. The astronomers Fred Hoyle and Chandra Wickramasinghe placed the probability that life would originate from nonlife as $10^{-40,000}$ and the probability of added complexity arising by mutations and natural selection very near this figure.[33]

But just what do these numbers mean? Evolutionists often respond that given the 5-billion-year history of the earth, such odds are not prohibitive. But compared to the probability of accumulating a long series of favorable mutants, 5 billion years is trivially short, and our universe is trivially small. There are "only" 10^{17} seconds in 5 billion years and 10^{80} atoms in the known universe. At one million events per atom per second for five billion years, only 10^{103} (10^6 x 10^{17} x 10^{80}) events could occur in all of earth's history. That is a large number compared to the number of events occurring in a person's lifetime, but infinitesimally small compared to the figures quoted above for random changes in DNA's base code (mutations) producing improved genetic information. Biologists usually flatly reject such figures. They claim that the figures and calculations do not fully explain living systems. The mathematicians are usually not deterred, however; but mathematicians and biologists both seem to miss the real issue. The question is not *how much time* is necessary for mutation and natural selection to bring about the current biological world, but whether mutations are *capable* of producing true biological novelty *at all*.

A Helpful Analogy

Mutations, remember, are mistakes, sometimes referred to as copying errors, analogous to errors in retyping a manuscript. The DNA in living cells contains coded information. It is not surprising that so many of the terms used in describing DNA and its functions are language terms. We speak of the genetic *code*. DNA is *transcribed* into RNA. RNA is *translated* into protein. Protein, in a sense, is coded in a foreign language from DNA. RNA could be said to be a dialect of DNA. Such designations are not simply convenient or just anthropomorphisms. They accurately describe the situation.

To further the analogy, the genetic code is composed of four letters (nucleotides), which are arranged into sixty-four words of three letters each (triplets or codons). These words are arranged in sequence to produce sentences (genes). Several related sentences are strung together and perform as paragraphs (operons). Tens or hundreds of paragraphs comprise chapters (chromosomes), and a full set of chapters contains all the necessary information for a readable book (organism). Like all other analogies, this one is not perfect, but the point is clear. One does not add constructive sentences, paragraphs, or chapters to a complete book by the selective addition of random copying errors. Is it therefore reasonable to expect evolutionary novelty to arise in living creatures by slow accumulation of point mutations?

A firm understanding of the complexity of DNA and its function has triggered remarks of incredulity by those contemplating the usefulness of mutations. Sir Ernst Chain, Nobel Prize winner in biology, wrote:

> To postulate that the development and survival of the fittest is entirely a consequence of chance mutations seems to me a hypothesis based on no evidence and irreconcilable with the facts. These classical evolutionary theories are a gross oversimplification of an immensely complex and intricate mass of facts, and it is amazing that they are swallowed so uncritically and readily, and for such a long time, by so many scientists without a murmur of protest.[34]

Most mutations obviously do not offer any great benefit. It is easy to see why most mutations are in fact disadvantageous. But suppose, just for a minute, a mutation occurs that does not harm the cell or organism. What can it really accomplish? A mutation in a structural gene will not produce anything new, just a minor variation of what already exists. A mutation simply rearranges what is already there.

The usual answer given to the dilemma of new genetic information is that as a gene continues to mutate, eventually something different will arise. But immediately several questions come to our minds. What function, for example, is this protein performing while all this mutating is going on? Is its function slowly changing? If so, is its former function still needed? If not, why not? And if so, then how is the former function being handled? How is the new gene escaping its old control mechanism and establishing new ones? But even beyond these questions, the only observational evidence we have so far contradicts the possibility that mutations can produce a truly new gene, a nonallele in a new category of structure and function. This is not to say that such evidence can never be found, but as each year goes by without such evidence, the likelihood diminishes.

What Is a New Gene?

It might be helpful at this point to draw a loose analogy to illustrate what we mean by a *new* gene. In a literal sense, any mutation produces a gene unlike the previous gene. It is no longer exactly the same. Therefore it is technically correct to say that a new and different gene has appeared. For example, sickle-cell hemoglobin is a new gene. But we are looking for something more than that. Such a mutational change is analogous to hiring a new teller at a bank. The same task is being performed but by a new person, and no two people do the same task exactly alike. But let's say that the bank decides to go into real estate. Banks don't simply train a loan officer to sell real estate. There is much more involved in such a decision. The change also requires new departments and new people with talents and abilities that are not currently present in the bank. In the same way truly *new* genetic information is not constituted by new *alleles* (alternate expressions of a gene) but by genes with a uniquely different structure and function.

The study of bacteria has been profoundly at the center of studies of mutations. This is because they reproduce rapidly, producing large populations and large numbers of mutants. They are also easily maintained, and their environments are easily manipulated in the laboratory. Despite all their advantages, never has there arisen in a colony of bacteria a bacterium with a primitive nucleus. Never has a bacterium in a colony of bacteria been observed to make a simple multicellular formation. Although hundreds of strains and varieties of *Escherichia coli* have been formed, it is still *Escherichia coli* and easily identified as such. In short, we must ask and conclude along with the French zoologist and evolutionist Pierre-Paul Grasse':

> What is the use of their unceasing mutations if they do not change? In sum, the mutations of bacteria and viruses are merely hereditary fluctuations around a median position; a swing to the right, a swing to the left, but no final evolutionary effect.[35]

It might be countered that these are asexually reproducing bacteria; what about eukaryotic, sexually reproducing, multicellular organisms? The best example is the fruit fly *Drosophila*, the genetic workhorse of the past sixty years. With a generation time of less than three weeks, in a two-year period, thirty to forty generations can be observed in laboratory-controlled conditions. A variety of means can also be used to produce new mutants or expose existing ones: selective breeding, chemical mutagens, and radiation. In other words, *Drosophila* has been subjected to numerous mutations in a variety of environments over periods of time covering many generations, an ideal test case for the Neo-Darwinian model. Yet despite the fact that *Drosophila* demonstrate a high degree of variability as mentioned in chapter 3 and can form new species readily, the fruit fly is always a fruit fly. Numerous recent authors have remarked at how morphologically static the genus *Drosophila* is.[36] Francis Hitching perhaps said it best: "Fruit flies refuse to become anything but fruit flies under any circumstances yet devised."[37] In response, it might be said that even with the shortened generation time and increased mutation rate, sixty years is a drop in the bucket of evolutionary time. Maybe so, but the experimental history has not so far verified the Darwinian expectations. Besides, placing one's hopes in the success of future experiments, when past experiments have had such a long negative track record, is hardly an enviable position for the Neo-Darwinists.

The enormity of the problem is realized when attention is shifted from the origin of a single biological novelty to the origin of the current diversity of life from a single one-celled ancestor. This requires more than just gene modification by mutation. A vast quantity of new genetic material is required.

> A gene cannot arise from just any other gene. The cytochrome *c* gene of man cannot arise by a mutation from the gene coding for hemoglobin in man, or from the gene coding for cytochrome *c* in wheat.[38]

Where Do We Get New Genetic Material?

So where does this new genetic material come from? A whole new functioning gene certainly does not simply appear as the result of a completely new sequence of nucleotides. The proposed solution, as stated in chapter 3, is gene duplication. This may mean duplication of a single gene, all or part of a chromosome, or the whole genome. Markert et al., in their 1975 paper "Evolution of a Gene," stated the concept well:

> Although genes evolve whether duplicated or not (for example, cytochrome *c*), the addition of new information requires duplication. In the complex organisms living today and for a long time into the past, the primary source of new genetic material (new information) must be duplicated copies of existing genes, since the creation of a new useful gene seems impossible.[39]

Stebbins adds:

> Just as all life has evolved from preexisting life, so all the functioning macromolecules that make up living systems have evolved from preexisting functional macromolecules.[40]

After a gene becomes duplicated, the ancestral gene continues its former function while the new gene is free to accumulate mutations that can eventually lead to a different function altogether. Independent evolution results. The most common example of gene duplication involves the globins: myoglobin and hemoglobin. Both myoglobin and hemoglobin are involved in oxygen transport, myoglobin in muscle and hemoglobin in blood. Myoglobin is a molecule composed of a single polypeptide (153 amino acids long in humans) twisted in a complex three-dimensional structure with a heme group. Hemoglobin contains four such polypeptides, two of one kind and two of another, each with its own heme group. In humans, there are four hemoglobin genes. Normal adult human comes in two forms, hemoglobin A, which consists of 2 α and 2 β or the rarer, hemoglobin A2, which consists of 2 α and 2 δ. Fetal hemoglobin is made up of 2 α and 2 γ

The evolution of myoglobins and hemoglobins is worked out as illustrated in Figure 19. A single myoglobin gene was the ancestral gene. There was allegedly a duplication around 650 million years ago in an ancestral vertebrate stock. The duplicated gene eventually became α-hemoglobin. Three hundred million years later, there was another duplication that gave rise to β-hemoglobin. With two hemoglobin genes present, there was now the possibility for the tetrameric structure of modern hemoglobin. There were then two more duplications that gave rise to the γ and δ genes respectively. Several other gene families have been related in a similar fashion.

present proteins:

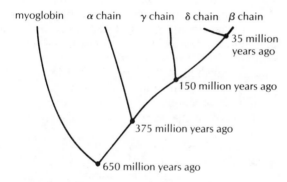

Figure 19. Conjectured Phylogenic Relationships. After Strickberger, Genetics, p. 841. fig. 19

But does this particular example and others like it demonstrate the mechanism for the origin of new genetic information? The answer appears to be negative, even assuming the above scenario to be true. After 650 million years of duplication and subsequent mutation, the various genes have not escaped their basic function of oxygen transport. Once an oxygen-transporting gene, always an oxygen-transporting gene. Does this duplication and subsequent mutation really constitute new genetic information? Once a gene performs a particular function, it is unable to completely depart from the structural and regulatory constraints placed on it. From the evidence available, we conclude that a dehydrogenase enzyme, for example, will either remain a dehydrogenase or lose its usefulness altogether. Cytochrome c will remain cytochrome c.

Possible exceptions are two isolated examples of similar proteins that perform different functions. Human insulin and mouse nerve-growth factor have similar amino acid sequences, though actually sharing only 17 percent (20/118 positions contain the same amino acid) of their residues. However, both elicit similar responses from their target cells. and show a similarity in their place of origin.[41] Also, α-lactalbumin, which is part of an enzyme used in the synthesis of lactose, and tear lysozyme, which degrades polysaccharides in bacterial cell walls, are similar. Nevertheless, here also there is a degree of similarity in the place of biological origin. There are also parallels in the type of substrate used by the enzymes.[42] With only these two examples, it is difficult to determine whether these two are similar proteins because of direct descent or not, though it remains a possibility.

In the case of hemoglobin, a further question is how the change from the monomeric myoglobin to the tetrameric hemoglobin was achieved. This is an upward step in complexity and metabolic efficiency. This would certainly require a change in the regulatory pathway as well as the duplication of the hemoglobin genes. In addition, human α, and β polypeptides share 43 percent (63 out of 148 proposed original positions) of their sequences. The β, γ, and δ are a much more tightly knit group. Human myoglobin and human hemoglobin share only 14 percent of their sequences. Therefore, the four hemoglobin polypeptides are more easily connected to each other than they are as a group, to myoglobin. The myoglobin-hemoglobin duplication is a much more tenuous proposal, making the jump from myoglobin to hemoglobin that much more difficult. Human beings undoubtedly contain thousands of unique genes not found in single-cell bacteria. Their origin is far from resolved in the Neo-Darwinian synthesis.

Frequently, upon receiving criticism about the sufficiency of point mutations, the Neo-Darwinist brings up the subject of natural selection. Certainly mutations are of no use without the creative guidance of natural selection. The examples of artificial selection are ushered forth along with a few examples from nature. However, even natural selection, the cornerstone of Darwinism, has been severely criticized by some evolutionists and abandoned altogether by others as an explanation of the origin of evolutionary novelty.

The problem usually revolves around Herbert Spencer's phrase "the survival of the fittest." Mere survival of the individual is no longer the criterion. It's how many offspring that individual will leave behind that is important. If an organism with a particular genotype continues to leave more offspring than other organisms do, its genotype will soon come to spread throughout the species.[43]

This all works out mathematically, since it is a circular truism or tautology.[44] Which are the most fit? Those that produce more successful offspring. But which are the ones that produce more offspring can only be deduced after the fact *because* they left more offspring. *Predicting* the outcome is practically impossible, as is trying to determine the process that led to the resulting change in gene frequencies. British evolutionist C.H. Waddington gave this effective summary:

> There you do come to what is, in effect, a vacuous statement: natural selection is that some things leave more offspring than others: and, you ask, which leave more offspring than others: and it is those that leave more offspring: and there is nothing more to it than that. The whole real guts of

evolution—which is how do you come to have horses and tigers and things—is outside the mathematical theory.[45]

Some, such as the paleontologist and evolutionary theorist Stephen Gould of Harvard, while admitting its misuse, have attempted to provide a more testable base.[46] Gould maintains that certain traits do offer advantages by the criterion of better design for living in new environments. These traits improve fitness by supplying more effective survival, not by the fact of their survival and spread. However, others feel that this fails to help because it is unable to provide a means to *predetermine* those traits that are better designed. D.E. Rosen from the American Museum of Natural History put it this way:

> Gould is unable to tell us how to recognize a successful organism apart from its existence, or what exactly is the reason for its existence as compared with others that have died off, or which attribute of the organism shall be compared with human engineering achievements, or even how such comparisons can be rendered objective.[47]

The problem is not that the process of natural selection does not take place in nature. It does, and it is an important process in nature. What we now call normalizing selection was known even before Darwin, though the phrase "natural selection" is original with him. The problem is simply that it is very difficult to predict which genotype will prevail based on objective criteria. Just as any Monday morning quarterback can offer a reasonable explanation for the previous day's result, so any trained biologist can offer an explanation for a change in gene frequency based on natural selection. But whether either explanation is accurate is nearly impossible to determine. Too many biological phenomena have been glossed over by simply saying "selection favored" this or that. Investigations of this sort need to dig deeper. Often a multiplicity of morphological, physiological, and behavioral factors will be involved. That a trait is selectively advantageous now does not *necessarily* mean that natural selection is the primary factor in its historical development.

The force of Darwin's argument for the importance of natural selection stemmed from his analogy to artificial selection in domesticated plants and animals. The analogy, however, breaks down on two counts. First, in artificial selection, there is a preconceived desired goal. A particular trait is designated for increase or decrease. This is true selection. In natural populations, this is not so. As was mentioned in chapter 3, natural selection is a passive phenomenon. There is no selector. No desired goal or sense of progress is possible. Natural selection operates on random mutations that do not anticipate a determined end result. Many have assigned a more powerful ability to natural selection than to artificial selection. This appears entirely out of order and unrealistic.

Second, as in the case of mutations and gene duplication, all the examples of artificial selection can be interpreted as demonstrating the opposite of the point to be made. Rather than showing that limitless change is possible, the overriding observation is that there are limits to change. Many varieties of chickens have been produced from the wild jungle fowl. But breeders have been unable to create any new varieties because all the genes in the wild jungle fowl have been sorted out into the existing varieties— there is strictly limited variation. Beginning in 1800, plant breeders sought to increase the sugar content of the sugar beet. They were successful. After seventy-five years of selective breeding, they were able to increase the sugar content from 6 percent to 17 percent. But no further progress could be made.

The examples in chapter 4 exhibit the same

results. Egg production was sharply increased but eventually leveled off. Oil and protein production in corn showing extensive variation because of the probable involvement of many genes never reached a limit. However, it seems that a limit would have been reached if the experiment had continued: the amount of change per generation was definitely leveling off. A rule that all breeders recognize, is that there are fixed limits to the amount of change that can be produced. It is of course unequivocal that there is extensive variation in nature, as documented in chapter 4. However, chickens don't produce cylindrical eggs. We can't produce a plum the size of a pea or a grapefruit. There are limits to how far we can go. An alternate explanation is that mutations may possibly narrow or broaden the limits, but they never break them. Some people grow as tall as seven feet, and some grow no taller than three; but none are over twelve feet or under two.

Artificial selection, then, is not the best analogy for natural selection. The two have little in common. Artificial selection demonstrates well the large amounts of genetic variation in nature, but it offers little in aiding our understanding of natural selection. This is particularly distressing as we attempt to understand nature's most perplexing questions, those concerning the extraordinary adaptations of living systems.

Criticism of Adaptation

The slow, gradual process of Neo-Darwinism requires that all adaptations be built up gradually with intermediate stages that are useful to the organism. This becomes taxing to the imagination when we consider some of the remarkable adaptations mentioned in chapter 2. Take the woodpecker, for example. How did its remarkable tongue begin its odyssey around the back of the head? How did its tongue make its initial break out of the throat cavity? What prompted woodpeckers to begin poking holes in dead trees? How did they know there were insects to be had in there in the first place? Similar questions could be asked about other adaptations in the woodpecker, such as the two-by-two toe structure and the shock-absorbing cartilage behind the beak. The question has always been, What good are these structures unless they are complete from the start? Stephen Gould asks the same question with a touch of humor:

> But how can a series of reasonable intermediate forms be constructed? Of what value could the first tiny step toward an eye be to its possessor? The dung-mimicking insect is well protected, but can there be any edge in looking only 5 percent like a turd?[48]

Complex adaptations have always provided the most rigorous test of Darwin's theory of natural selection. Every naturalist has his own favorite. Darwin's was the eye. Much has been made of Darwin's befuddlement over the eye. His exasperation is expressed in *Origin of Species* and in several letters to colleagues. It certainly seems unreasonable to propose that the eye developed slowly by chance mistakes to its present complexity of nerves, cones, rods, lens, pupil, retina, etc. If the eye is not capable of providing a picture that successfully aids in survival, of what good is it? If all the parts are not in sufficient working order, it is of no use to the organism. The more we learn of the eye, the more we appreciate Darwin's fear. Darwin's only suggestion, which remains today, is to observe the many types of eyes found in lower creatures, from the light-sensitive structures on some single-celled organisms to the compound eye of insects, to the focusing, color-receiving human eye. This is satisfactory to most Neo-Darwinians, but does it actually solve the problem? All that this technically demonstrates is that there are various organisms that possess light sensitivity in nature. Each contains its own complexities and unique properties. No true *historical* series has ever been proposed for the evolution of the human eye. Darwin's solution may only add new examples to the list of complex adaptations waiting to be explained.[49]

When dealing with complex adaptations, Neo-Darwinians use the term *preadaptation*. They admit that incipient structures did not perform their future function throughout their development. While performing a different function, the structure was becoming prepared or was preadapted to perform another function. It should be understood that no concept of goal-oriented evolution is ever implied. The bones in fish jaws, for example, are said to have been present in ancestors but supported the gill arch and thus were functioning in a respiratory role, though "preadapted" for eating.

This kind of explanation may be helpful in describing the origin of adaptations such as bird feathers. Bird feathers are constructed uniquely to aid in flight. The theory suggests that as their finely tuned structure evolved from scales, their function was that of heat insulation. In order to aid in flight, a complex structure was required that had to develop in a step-by-step fashion. In the meantime, feathers merely provided an insulating layer. Though the reasoning is enticing, the conundrum remains. For though bird feathers have an integrated structure of their own, they are only one aspect of the coordinated and complex adaptation of bird flight that requires neurological, skeletal, and muscular adjustments as well. In light of our previous discussion of mutations, we are still in the dark as to how the single transition from scale to feather, to mention only one necessary change, could have been effected by mutation events. The evolution of other coordinated structures such as the jaw, ear, and eye is not really addressed. Simply observing similar bones in another organism performing a different function does not tell us how the old function was relinquished and the new one assumed. How were the bones supporting the gill arches allowed to give up supporting gill arches and begin acting as jaw bones? In another case, simply moving the bones from the reptilian jaw into the mammalian ear does not automatically mean these bones are capable of aiding the coordinated function of hearing. They were just bones in the ear.

Did Incipient Organs Have Other Functions?

To summarize the problem, if incipient organs or structures were performing a different function in their preadapted state, why or how did the organ switch from this function to the one for which it is preadapted? And what took over the incipient organ's previous function? Was it lost or was it also mysteriously taken over by another preadapted structure? To be sure, the problem of perfection in highly complex organs that has plagued Darwinism from the beginning has in no way been resolved.

Environmental Tracking

That natural selection is capable of providing a mechanism for producing evolutionary novelties or adaptations is far from being established. Natural selection, therefore, is increasingly being returned to its pre-Darwinian status as a conserving mechanism. This explanation is sometimes referred to as environmental tracking. The concept of differential reproduction insures that an organism will be able to keep up with its ever-changing environment. If the necessary genetic variation is present, tracking the environment or fine tuning itself allows the species to persist within the limits of its available hereditary variability. Environmental tracking provides solely for maintaining an organism's adaptive capabilities, not for introducing totally new solutions to new and yet unsolved environmental challenges.

The problem with the theory of environmental tracking is that it does not predict or explain what is most dramatic in evolution: The immense diversification of organisms that has accompanied, for example, the occupation of land from the water or of the air from the land.[50]

The article by R.C. Lewontin from which we have just quoted contains an excellent analysis of current evolutionary thought about adaptation. Following the above statement of the problem, Lewontin explains that the role of the evolutionary biologist is to "construct a plausible argument about how each part (or an organism) functions as an adaptive device." It is not surprising to note that the task is to describe the adaptation, not how it evolved. It is simply *assumed* that it was molded by natural selection. This appears to put the cart before the horse, especially in light of his statement on environmental tracking. This is said to be resolved, however, by applying an engineering analysis to the adaptation and its environment. When this analysis combines with the principle of the struggle for existence, one has the capability of predicting which of two organisms will leave more offspring (removing perhaps the apparent tautology).

> An engineering analysis can determine which of two forms of zebra can run faster and so can more easily escape predators; that form will leave more offspring. An analysis might predict the eventual evolution of zebra locomotion even in the absence of existing differences among individuals, since a careful engineer might think of small improvements in design that would give a zebra greater speed.[51]

Lewontin rightly points out, however, that this could be very misleading because there may be other side effects to running faster, detrimental to the zebra, that could not be predicted. The organism's life history must be nearly completely understood before such analysis could be done with any degree of confidence. Besides, the dilemma is not really resolved, because all that is really being discussed, at least in this example, is a faster zebra, not a truly novel adaptation such as "legs" on a fish preadapted to land.

In addition to the problem of isolating a particular adaption from the rest of the organism for analysis, Lewontin lists several other processes that can influence the morphology, physiology, and behavior of an organism that have nothing to do with adaptive processes and thus confuse the issue even more. After reflecting extensively on the problems and alternatives of the adaptationist argument, Lewontin asks why it is still pursued. His answers deserve close attention.

> On the one hand, even if the assertion of universal adaptation is difficult to test because simplifying assumptions and ingenious explanations can almost always result in an ad hoc adaptive explanation, at least in principle some of the assumptions can be tested in some cases.[52]

Lewontin goes on to say that the other alternatives are also testable only some of the time. The temptation, then, is to explain those adaptations that lend themselves to testing and leave the difficult ones to chance. Surprisingly, then, he concludes that the adaptationist program is forced on the biologist because the other alternatives, though certainly operative in some cases, are untestable in many cases. In other words, the adaptationist argument is "forced" on biologists because, though its assumptions are only occasionally testable, so are the alternatives only occasionally testable. Where is the compelling logic? Perhaps it is in the second reason.

Lewontin states that to throw out the adaptationist program would be a mistake because adaptation is a real phenomenon:

> It is no accident that fish have fins, that seals and whales have flippers and flukes, that penguins have paddles and that even sea snakes have become

laterally flattened. The problem of locomotion in an aquatic environment is a real problem that has been solved by many totally unrelated evolutionary lines in much the same way. Therefore, it must be feasible to make adaptive arguments about swimming appendages.[53]

The logic is elusive here as well. To state that a particular course is correct simply because it has to be is hardly compelling. A theory needs to be able to stand on its own merits, not because other ideas don't measure up. The adaptationist program is continued mainly because Neo-Darwinists are convinced it is true, never mind that the reasons for this conviction are illusory. No one really questions that fish, seals, whales, penguins, and sea snakes are "adapted" to travel in water. The question is, How did mutation, guided by natural selection, arrive at these similar solutions? That it did is merely assumed and never satisfactorily explained.

The Relevance of Population Genetics

One of the ways in which a population geneticist will explain evolution is to define it as changing gene frequencies. Though this is not what most people understand evolution to be, it is what population geneticists study. Population genetics is the study of the relationship through time of two or more alleles of a given locus by observing their phenotypes in natural or laboratory populations. Much of the support for Neo-Darwinism in the past few decades has come from the work of population geneticists. It is of no surprise, then, that the main defendants in the recent attacks on Neo-Darwinism have been population geneticists.

The major limitation of population genetics is time. Since population geneticists deal with real, observable populations, the most they can observe is tens of generations over a few years. This becomes a mere water molecule of the proverbial drop in a bucket of proposed evolutionary time. Four of the five cases presented in chapter 4 for mutation and natural selection are merely examples of fine-tuning or environmental tracking. The fifth case, sickle-cell anemia, is unique; it shows little possibility of continuing evolution.

In bacterial antibiotic resistance, the mutation is, first of all, a recurring one. This is demonstrated by the repeatability of the experiment. Also, the rate of occurrence was overestimated because the mutation was spread rapidly through gene transfer between cells, giving the appearance of many individual mutants.[54] In some cases, the mutation drastically altered the normal functions of the cell, producing an "evolutionary cripple."[55] The bottom line, however, is that this favorable mutation simply allows the organism to continue its role and persist as a species.

Insect pesticide resistance falls ito a similar category. In this situation, however, it is impossible to determine whether the resistance is due to a new mutatioin or to variation already present in the population. Pesticide resistance has taken various forms, including decreased uptake of the pesticide and an increased ability to detoxify either by modifying an existing enzyme posttranscriptionally (the gene then being unaffected) or by altering the amount of the enzyme produced.[56] The end result is that in each instance there was an intensification of a faculty the insect already possessed, a classic example of environmental tracking.[57] Therefore, in both bacterial antibiotic resistance and pesticide resistance. The populations possessed the enzymes and metabolic processes at a low activity before exposure. The antibiotic or pesticide simply acted as a selecting agent to weed out those that were less efficient. This is evolution in the sense that there was a change in gene frequency, but not in the sense that the organism obtained a totally new cellular function.

The famed peppered moth is even less impressive because the melanic form appears to have always been present in small numbers.[58] This example, then, becomes an example of shifting gene frequencies in response to changes in the environment that are now reverting back to the initial condition. With the risk of being redundant, *Biston betularia* remains *Biston betularia*, fine-tuning at its best.

The experiments with the irradiated fruit flies offer a more interesting challenge. Here is a case in which a mutation is generated that increases fitness. The first point, as in all the other cases, is that we are dealing merely with a more efficient fruit fly. Here, evolution has occurred in the sense of fine-tuning, but not in the major sense. Also, that a favorable mutation has been generated is not certain. It is merely *assumed* that the increased carrying capacity and faster reproductive rate is due to favorable mutations. No alternatives were explored. One possibility is that those flies with weaker resistance to mutation from any source were wiped out of the experimental population, leaving a hardier population behind. The control population remained saddled with individuals of lower tolerance to mutations and, therefore, lower reproductive rates. Selection still occurs, but the emphasis shifts. The experiment becomes the removal of the unfit exposed by the *deleterious* mutations due to irradiation, rather than an increase in general fitness caused by *favorable* gene mutations.

The case of sickle-cell anemia is the poorest of examples. The sickle-cell mutant allele has no chance of eliminating the normal allele because it is lethal in the homozygous state. Sickle-cell anemia is a useful example of balancing selection, but serves little purpose in illuminating the cause of major biological change.

Another major limit for population geneticists is determining the meaning of the genetic variation measured by gel electrophoresis and other methods. One problem is the lack of certainty in knowing how much variation is actually being measured by gel electrophoresis. Since electrophoresis measures primarily changes in the charge of the protein, only about one- to two-thirds of all amino acid substitutions would be detected. This cryptic variation has been verified by other biochemical tests and, finally, by peptide mapping. These studies have suggested that the level of cryptic variation has been overestimated, yet is significant enough to increase estimates of average heterozygosity by two or three times in some organisms.[59] Ambiguity in this area is compounded by differences in the standards of interpretation of banding patterns, lack of uniformity of loci studied, and too few loci studied to render conclusions that are comparable and truly representative.[60] All of this indicates that these data must be viewed tentatively. Far from useless, the data are, however, still open to a variety of conclusions.

Problems in assessing the true significance of gene-protein polymorphism are evident in the longstanding controversy over the neutral or adaptive nature of protein variation. This stems primarily from the difficulty in correlating levels of protein variation with factors from the environment. It was initially thought that all protein variation was in some way significant to the organism. High levels of variation were an adaptation for surviving in heterogeneous environments. In the late 1960s, however, some began to suggest that most, if not all, of the protein variation was due to neutral mutations whose frequencies were due solely to random drift.[61] Though the two camps were initially extremely polarized, there is an increasing awareness that the truth lies somewhere in the middle. There is a continuum from lethal to neutral to adaptive, and protein variants fall somewhere on the line, not at one end or the other nor predominantly in the middle.[62] Neo-Darwinists in particular, however, stubbornly hold onto the notion that all genetic variation at some time in its history was favored by natural selection, though neutral at present.[63]

The point of this particular discussion is to clarify the significance of population genetics. Its data are frequently used to bolster the Neo-Darwinist position. Our argument, and it is not ours only, is that the data on protein variation are able to shed only a dim light on the major problem of evolution—the appearance of novel adaptations. The major significance of population genetics is in helping us to understand how an organism responds to minor environmental fluctuations. And even this can be clouded in fundamental differences in theory.

Criticism From Paleontology

In the last decade, major critics of the Neo-Darwinian synthesis have arisen from the ranks of paleontologists. Their criticisms are primarily twofold. First, whenever there is an opportunity to observe an organism through successive periods of geologic time, the species shows little or no morphologic change. This phenomenon is referred to as stasis and can cover hundreds of millions of years. Stephen Stanley, a paleontologist from Johns Hopkins, gives a number of examples of "living fossils." A few with their lengths of duration in millions of years are listed below.[64]

Sea urchins	230 MY
Horseshoe crabs	230 MY
Bowfins	105 MY
Sturgeons	80 MY
Bats	50 MY
Alligators	35 MY

This evidence is seen as contradictory to Neo-Darwinism or gradual evolution for two reasons. First, the origin of these animals is sudden and seemingly complete. The earliest horseshoe crab found in the fossil record is nearly identical to the horseshoe crab that exists today. Second, although there have been numerous environmental fluctuations, these animals remain virtually unchanged, even to the present day. Gradual evolution would have produced continuous change, especially over tens of millions of years. This would not be particularly persuasive if stasis were an exception to the rule, but to punctuationalists such as Stanley, it *is* the rule. In just about all cases, once a species appears, it remains constant morphologically and either persists to the present day or becomes extinct, but no major evolutionary transitions are observed.

In addition to stasis, the other objection to gradual evolution is the lack of transitions between species and major new groups of organisms. The lack of transitions or "gaps in the fossil record" has been discussed openly over the last decade, so there is little need to elaborate, except to quote a number of paleontologists and Neo-Darwinian critics from the past few years.

David Kitts, 1974:

Despite the bright promise that paleontology provides a means of "seeing" evolution, it has presented some nasty difficulties for evolutionists, the most notorious of which is the presence of "gaps" in the fossil record. Evolution requires intermediate forms between species and paleontology does not provide them.[65]

Anderson and Coffin, 1977:

The search for these transitional forms (missing links) by paleontologists has not been very successful. Each major group of organisms appears abruptly in the fossil record without any transitions.[66]

David Raup, 1979:

Instead of finding the gradual unfolding of life, what geologists of Darwin's time and geologists of the present day actually find is a highly uneven or jerky record; that is, species appear in the sequence very suddenly, show little or no change during their existence in the record, then abruptly go out of the record.[67]

Stephen Stanley, 1979:

The known fossil record fails to document a single example of phyletic evolution accomplishing a major morphologic transition and hence offers no evidence that the gradualistic model can be valid.[68]

Stephen Gould, 1980:

The extreme rarity of transitional forms in the fossil record persists as the trade secret of paleontology. The evolutionary trees that adorn our textbooks have data only at the tips and nodes of their branches; the rest is inference, however reasonable, not the evidence of fossils. All paleontologists know that the fossil record contains precious little in the way of intermediate forms; transitions between major groups are characteristically abrupt.[69]

Francis Hitching, 1982:

When you look for links between major groups of animals, they simply aren't there; at least, not in enough numbers to put their status beyond doubt. Either they don't exist at all, or they are so rare that endless argument goes on about whether a particular fossil is, or isn't, or might be, transitional between this group and that.[70]

The list of difficulties with the Neo-Darwinian interpretation of evolution is indeed long. The apparent insufficiency of mutations, the lack of an adequate explanation for new genes with new functions, the ambiguity of adaptation and natural selection and their use as testable theories, the difficulty in extrapolating from population genetics to major evolutionary trends, and finally the paleontological inconsistencies should render Neo-Darwinism (or the Modern Synthesis) as a very tentative hypothesis. Indeed, many writers, though almost exclusively evolutionists of one sort or another, referred to in this chapter have declared it to be dead.

But scientists are human too, with attachments to theories that are maintained as much by predetermined bias as by the facts. Most, if not all, of the elements of Neo-Darwinism will continue in popularity for some time. This is evidenced in the summary book on evolution by G. L. Stebbins.[71] Though modern methods and data are included with a few new twists, the basics are still the same. Point mutations, natural selection, and extrapolation from microevolution to macroevolution is all that is really necessary to understand evolution.

Others, however, are not satisfied with merely giving the Modern Synthesis a facelift. There is a strong movement, principally among paleontologists, to evoke a major shift in emphasis from gradual evolution to evolution that is episodic in nature. We will now turn our attention to this new theory of punctuated equilibrium.

chapter six

Evolution by Speciation

Punctuated equilibrium has arisen as a more inclusive evolutionary theory of biological change.

With criticisms of Neo-Darwinism's major tenets mounting, it is no surprise to find alternative evolutionary ideas coming forward. In 1972 Niles Eldredge and Stephen Gould[72] proposed the concept of punctuated equilibrium as just such a theory. Their thesis was based on two primary paleontological observations. The first is the presence of gaps in the fossil record between species and higher taxonomic categories. In the previous chapter, this phenomenon was cited as a critical problem in the Neo-Darwinian model. Second, and more important, is the observation that once a species appears in the fossil record, its morphology changes to only a trifling degree.

111

From these two observations, Eldredge and Gould postulated that on a geological time scale, new species arose with sufficient suddenness as to appear instantaneously in the fossil record. *This would account for the gaps.* Once in existence, the species would stabilize, adjusting only to minor environmental fluctuations until it experienced the ultimate fate of all species, extinction, virtually unchanged. Evolution would be episodic rather than gradual in the Neo-Darwinian sense. The term *punctuated equilibrium,* then, is easily explained. For 99 percent of a species' existence, it survives at an equilibrium, with minor fluctuations. This equilibrium, or period of stasis, is punctuated by a rapid speciation event. The new species eventually settles down to a new and different period of stasis.

At this early stage, it is critical to make an important distinction. Technically, punctuated equilibrium is a description and not necessarily a mechanism or process. It is an attempt to describe the data of paleontology at their face value. Just what genetic mechanism is involved in the speciation process was not initially explored. Consequently the genetic aspects of this new development are open to a wide number of alternatives. With Neo-Darwinism, the gradual accumulation of point mutations over long periods of time was easily identified as the mechanism. No such clear-cut mechanism is available for scrutiny with punctuated equilibrium.

But before examining the possible mechanisms, it is important to discuss the relevant data from the fossil record that brought about the hypothesis of punctuated equilibrium.

The history of most fossil species includes two features particularly inconsistent with gradualism:

1. Stasis. Most species exhibit no directional change during their tenure on earth. They appear in the fossil record looking much the same as when

they disappear; morphological change is usually limited and directionless.

2. Sudden appearance. In any local area, a species does not arise gradually by the steady transformation of its ancestors; it appears all at once and "fully formed."[73]

Gould in a later article[74] carefully emphasized that of the two claims of punctuated equilibrium as stated above, stasis is the most important. The other claim, sudden appearance, was made because of the famed gaps in the fossil record. Missing evidence proves difficult to quantify and study. Stasis on the other hand is real data. Gould goes so far as to say that "the potential validation of punctuated equilibrium will rely primarily upon the documentation of stasis."[75]

Stephen Stanley has marshaled the most extensive documentation of stasis to date.[76] Most pertinent is the discussion of "living fossils."[77] Living fossils are classified as those that are (1) still alive today, (2) have survived for long intervals of geological time, and (3) exhibit primitive morphologic characters similar or identical to those present when they first appeared. Stanley feels that living fossils offer a unique evidence for the punctuational model over gradualism. With few speciation events being discernible in these lineages, there was little opportunity for any further evolutionary change. (It is within speciation that punctuationalists describe most evolutionary change, as we will see later.) This is so, despite the wide diversity of environmental changes that were experienced during the long duration of their existence. Neo-Darwinism or gradualism would have predicted that with mutations continuously occurring, the organism would have evolved in response to environmental pressures. The difficulty Neo-Darwinism has experienced with these living fossils is exemplified by the divergent explanations. Some feel they have not evolved because they are very narrowly adapted to a specialized existence. Their microhabitats have persisted.[78] Others have proposed that they are very broadly adapted to a wide range of habitats and would therefore experience little evolutionary change through time.[79] Listed below are a few of the more notable living fossils, with longevity listed in millions of years (MY).

Horseshoe Crabs	230 MY
Bairdiid Ortracods	230 MY
Notostracan Crustaceans	305 MY
Bowfin Fishes	105 MY
Sturgeons, Gars, Sirens	80 MY
Snapping Turtles	57 MY
Alligators	35 MY
Aardvarks	20 MY

The punctuational model, then, predicts that even the persistent small groups will tend to exhibit little evolution because, however long they may have existed, they have undergone little speciation, and it is in speciation that most evolution occurs.[80]

The second aspect of punctuated equilibrium, sudden appearance of new forms, has been introduced already in chapter 5. We pointed out that the gaps in the fossil record are systematic. The punctuationalists claim these gaps can no longer be attributed to the incompleteness of the fossil record. This is most obvious in the sudden appearance of large numbers of diverse taxa seemingly at the same time. There are at least three of these occurrences in the fossil record: the Cambrian explosion, the appearance of the angiosperms (flowering plants), and the mammals.

Adaptive Radiation

Beginning with the Cambrian explosion, punctuationalists such as Stanley[81] and Gould[82] argue that there is now sufficient fossil documentation to indicate that this phenomenon, long an evolutionary enigma, is explainable in terms of adaptive radiation. Adaptive radiation on a small scale is a familiar concept to ecologists; it is defined as the opening up of new adaptive zones that makes possible the evolution of new and diverse species. The classic example has always been Darwin's finches on the Galapagos Islands. In this case, one, at the most two, species of finches arrived on the unpopulated, recently formed Galapagos Islands and eventually radiated to thirteen rather diverse species.

This principle is applied in a similar vein to three paleontological cases: the Cambrian explosion, the appearance of the angiosperms, and the mammalian radiations. A new organism that is able to take advantage of a new adaptive zone evolves and quickly radiates to fill it with new diverse species. Stanley reports in regard to these three examples.

Cambrian:

In summary, what has unfolded with the many discoveries of the past two or three decades is a picture of rapid, but not instantaneous, diversification of life during the latest Precambrian and Early Cambrian—an interval in the order of 100 million years.[83]

Angiosperms:

What the record reveals is that considerable diversification from these early forms took place during just 10 million years or so. This is not long by geological standards, but neither is it instantaneous. Darwin's abominable mystery is soluble, but here, as for the adaptive radiation of Cambrian marine life, the brief timescale opposed gradualism.[84]

Mammals:

Their great adaptive radiation was recent enough that the fossil evidence for it is impressive. Within perhaps twelve million years, most of the living orders of mammals were in existence, all having descended from simple, diminutive animals that might be thought of as resembling rodents, though not all possessed front teeth specialized for gnawing.[85]

Stanley proceeds to point out, particularly in regard to the mammals, that there is nowhere near sufficient time for the gradualistic model to account for such rapid diversification. The punctuation model with its many speciation events purports to offer a more plausible explanation of, or at least descriptive scenario for, these three remarkable phenomena.

Species Selection

As noted earlier in the section on living fossils, speciation is the primary agent for large-scale evolutionary change according to punctuated equilibrium. If a taxonomic group does not speciate, then there will be little change even though the lineage persists for extensive periods of time. Those taxa that are able to speciate provide new species units to test out the new adaptive zones. Some persist. Others go extinct. This has led to a concept known as species selection. It is formulated as analogous to natural selection among individuals. The species corresponds to the individual in natural selection. Speciation and extinction are like the birth and death of an individual. The speciation process is also analogous to mutation and recombination in that it provides for variability. Species selection will then operate like natural selection. Each new species is like the birth of a new individual with a unique genetic complement. Each species experiments with the environment with unique capabilities for survival. The better a taxon is able to speciate, the better its chances of surviving environmental fluctuations.

With this foundation, there are essentially three major sources of macroevolutionary trends, according to Stanley, phylogenetic drift (analogous to genetic drift), directed speciation (analogous to mutation pressure), and species selection.[86] Phylogenetic drift produces trends predominantly within small clades,[87] a clade being a cluster of lineages arising from a single lineage. Genetic drift is the establishment of certain alleles due to random sampling of the gene pool. In phylogenetic drift, patterns of speciation and extinction are more heavily influenced by random factors. Overall, phylogenetic drift plays a minor role, as does directed speciation. Directed speciation is observed when "a trend will form if every step of a single chain of speciation events moves in the same adaptive direction."[88]

By process of elimination, species selection is left as the dominant force in macroevolution. But even though the term *selection* appears, there exists a strong random element even here. The speciation event is likened to a mutation representing a dead end or perhaps the opening of a new adaptive zone. With the emphasis of species formation being on small isolated populations, two other factors contribute to the randomness of the process. The first is the founder principle, which describes a small local isolate population that embodies a nonrepresentative sample of the parent gene pool. The second is genetic drift, which can produce substantial changes in small populations by consequence of the random shuffling of genes through independent assortment, recombination, and the "luck of the draw" in the gametes that comprise the new individuals.

The primary selective agent in species selection was originally described as being contained within differential extinction. However, this ran contrary to natural selection, which is characterized by differential reproduction. Consequently Gould has acknowledged that differential speciation, the ability of one portion of a clade to speciate more rapidly than the other, will also lead to large-scale trends.[89] Under the banner of species selection, there are, therefore, two modes. In differential speciation, one portion of the clade speciates more rapidly, shifting the direction to one side. In differential extinction, most speciation events in one direction become extinct sooner, giving the same overall effect (Figure 20).

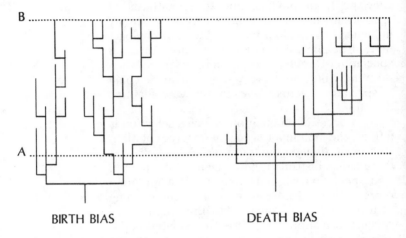

B

A

BIRTH BIAS   DEATH BIAS

Figure 20. Punctuated Equilibrium: Explanations of Evolutionary Trends. We begin at time A with two kinds of species, each of equal number within the clade. At time B, the descendants of both kinds remain at the ancestral mode, but differentail speciation permits one kind to dominate the clade. Right. death bias. Most species give rise to two descendants, one in each direction. But species to the right live longer than species to the left, thus helping to power the trend. Birth and death biases are modes of species sorting origin bias is the macroevolutionary analog of mutation pressure. After R. Milkman, ed., Perspectives on Evolution *(Sunderland, Mass.: Sinauer Assoc., 1982), p. 93.*

The delineation of species selection as a distinct process is perhaps the most significant aspect of punctuated equilibrium in contrast to Neo-Darwinism. Species selection effectively decouples macroevolution from microevolution. According to Gould,

> This higher order sorting of species, produced by differential origin and extinction, must direct evolutionary trends within clades [macroevolution] just as natural selection, acting by differential birth and death of bodies, directs evolutionary change within populations [microevolution].[90]

Neo-Darwinism subsists on the extrapolation of microevolutionary processes to account for macroevolutionary change. Species selection introduces a hierarchy to evolutionary processes. Various levels of the hierarchy, however, are not exclusively independent. Rather, there is hierarchy in the sense that the common body of causes and constraints act in characteristically different ways on the different levels.[91]Selection on the level of the individual has only an indirect effect on species selection. And neither level is more important than the other. Microevolution powered by natural selection produces the fine-tuning to minor environmental fluctuations. Macroevolution powered by species selection produces the large-scale trends documented in the fossil record.

Now we must turn our attention to the genetic mechanisms that are compatible with punctuated equilibrium. After all, true evolution cannot occur without genetic alterations of some sort. Of first priority is the genetics of speciation. We must discover if current biological concepts of speciation are compatible with the versions of the punctuationalists.

Is the process of speciation sufficiently rapid to expect no transitions in the fossil record? Indeed, what, if anything, do we know about speciation? How is it accomplished? What factors are important? Second, what sorts of genetic mutations are necessary to bring about rapid evolutionary change? If point mutations are insufficient, then what else is available? To establish any sort of heritable change, the genetic structure must be altered. These questions have all been addressed in the last few years but not always in connection with the theory of punctuated equilibrium. We will attempt to deal only with those aspects that have been discussed by punctuationalists or are, at least in our judgment, easily encompassed by them.

To understand speciation, we must first understand what we mean by a species. This is not as easy a task as it first appears. Some have argued that since a species is constantly changing, there is no such thing as a real species unit. However, the orthodox view of population biologists today is that in sexually reproducing organisms, the species is as real and almost as distinct as the individuals that comprise it.[92] The clearest and most accepted definition of a species was offered by Ernst Mayr: "Species are groups of interbreeding natural populations that are reproductively isolated from other such groups."[93] Although this appears simple enough, the trouble is that in nature the issue is rarely black and white but is made up of interminable shades of gray. There are degrees of reproductive isolation, both in the laboratory and in nature. This is to be expected if speciation is to be labeled a process. One is bound to run across groups of populations that are in process.

With the reference to reproductive isolation, we have quickly introduced the central question of speciation. How do two populations, once of the same species, become reproductively isolated so as to form two separate species? The prevailing concept of Neo-Darwinism was the allopatric model. In this model, two populations become isolated from each other by a geographic barrier that interrupts gene flow between the two populations. They can no longer come in contact with each other to interbreed. Some examples of a geographic barrier may be the formation of a new mountain range, the changing of the course of a river, or the interruption of prairie by stands of forest or vice versa. Any geographic or ecological change that would inhibit contact between the populations is a potential isolating barrier. With the onset of geographic isolation, the two populations could begin to experience subtle differences in their environment. As each population independently "tracks" its environment, phenotypic and genotypic differences would come about. As time passed, this divergence would continue, each population adapting to a slightly different environment. At some

point the geographic barrier may break down, or
it may be "punctured," allowing contact between the two populations.

Now, even though reproductive isolating
mechanisms may be starting to take effect,[94] the
two populations may be yet capable of interbreeding. However, it would not be to their
advantage to do so. Each population is now
adapted to its own particular environment. To
interbreed with the new population would place
the offspring at a disadvantage in either environment, since its genotype would be comprised of
elements of both. It is at this point that the
establishment of reproductive isolating mechanisms would be advantageous or adaptive.
Those individuals are selected that for one
reason or another are unable to interbreed with
members of the sister population and are therefore able to maintain the integrity of their own
population's gene pool. These individuals produce greater numbers of successful offspring and
are favored by natural selection.

It becomes clear that speciation in the Neo-
Darwinian tradition is, from start to finish, an
adaptive event. Although there are numerous
variations, most biologists until recently understood speciation in terms of the allopatric model.
Other forms of speciation were quickly dismissed for one reason or another. Recently,
however, various modes of animal speciation
have gained in respectability.[95] This has opened
the door for the consideration of types of
speciation that are more rapid and not so heavily
tied down to the slow, gradual, and adaptive
pace of the classic allopatric model. This change
is viewed as essential to the punctuationalists'
proposals.

The major emphasis of the punctuationalists,
however, has been on a variation of the allopatric model. Speciation by founder effect has
always been recognized as an option but has
never been given a strategic place in the Neo-
Darwinian synthesis. Indeed, it is Ernst Mayr

who cogently outlines this type of speciation, and it is Mayr who is extensively quoted by punctuationalists as having anticipated their views. [96]

Mayr defines the founder principle as "the establishment of a new population by a few original founders (in an extreme case, by a single fertilized female) that carry only a small fraction of the total genetic variation of the parental population." [97] With only a small amount of the initial variation present, the likelihood of this sample being a nonrepresentative sample of the parental population looms very large. Since the founder population is defined as being very small, certain alleles can be quickly eliminated because of genetic drift, inbreeding, and changes in the selection values of some alleles further reducing the genetic variability. This sets up what has been termed a genetic revolution in the speciating population. [98] The gene pool of the new population is now not only very different from the original parental stock but also very uniform. Very little intrapopulational variation goes on.

This would seem to provide just what the punctuationalist ordered. There appears to be the possibility not only of rapid speciation, but also of rapid genetic change that would presumably also lead to morphologic change. There are also other modes of speciation that can occur rapidly—parapatric and sympatric. Both parapatric and sympatric modes involve populations that are still in spatial contact with each other, either with similar populations as in the parapatric mode, or with the major parent population as in the sympatric. Only recently has the dust been blown off these two formerly discarded speciation modes, both of which may occur rapidly to meet the requirement of punctuated equilibrium.

Rapid Speciation

In searching for examples of rapid speciation among living species, Stanley offers three major criteria.[99]

1. The process of divergence must have begun only a short time ago (one thousand to ten thousand years ago). This would make it amenable to dating.

2. The divergent population must now be reproductively isolated, a situation that in many cases is testable.

3. There must be clear evidence of substantial morphologic divergence, a quantifiable quantum step.

Of these three criteria, the first two are perhaps the easiest to demonstrate. It is the third that is ambiguous and open to question. As Stanley readily admits, "We have so few excellent examples that we are confined to a piecemeal approach."[100]

Stanley,[101] as well as Mayr,[102] lists a number of examples of rapid speciation. On the island of Newfoundland, for instance, there are fourteen species of mammals, ten of which have diverged to the level of subspecies in the last twelve thousand years since the island became habitable with the retreat of the ice sheets. One species in particular, the Newfoundland beaver (*Castor canadensis caecator*), has been elevated by many as a separate species. There are several populations of the house mouse (*Mus musculus*) that have shown marked divergence in less than one hundred to three hundred years on islands in historical times. On Hawaii there are several species of moths that feed exclusively on bananas, though other Hawaiian species of this genus (*Hedylipta*) feed on a wide variety of food sources. Bananas were introduced to the islands by the Polynesians only about a thousand years ago; thus the former species of moths have become adapted to a sole diet of bananas in only this thousand-year period.

One other major example involved the cichlid fishes of Lake Nabugabo, a small lake bordering

Lake Victoria in Uganda. The lake was formed by the growth of a sandpit across an embayment in Lake Victoria an estimated four thousand years ago. There are six species of cichlids that inhabit the lake, one of which is indistinguishable from species in Lake Victoria whereas the other five are endemic. Although the five species show definite relationships to species in Lake Victoria, they differ in male coloration and other minor traits. Male coloration is the most significant difference, since it is used as a recognition factor in breeding. But neither this example nor the others previously mentioned meet Stanley's third criterion of substantial morphologic change. For evidence of this kind, he turns to several examples, only three of which we will discuss: the pupfishes of Death Valley, the cichlid fishes of Lake Victoria, and the honeycreepers of Hawaii.[103]

Death Valley, on the California-Nevada border, contains numerous isolated pools and streams inhabited by relic populations of fishes of which four species of the pupfishes (*Cyprinodon*) are only one group. It is estimated that the region became a desert following the retreat of glaciation twenty to thirty thousand years ago. *Cyprinodon milleri,* discovered in 1967, inhabits an area said to be only a few thousand years old. Its distinctiveness lies in its teeth and the near lack of pelvic fins. The most distinctive, however, appears to be the devil's pupfish (*Cyprinodon diabolis*), so named for its partiality to 92°F water. It lives solely in one small thermal spring, and its population probably never rises above a few hundred individuals. The devil's pupfish, aside from its ecological distinctives, is also the smallest by far of the four species, has reduced or no pelvic fins, and does not clearly resemble any of the other three species.

Lake Victoria, the parent body of Lake Nabugabo mentioned earlier, is estimated to be 750,000 years old, yet the 170 species of cichlid fishes that inhabit the lake are found nowhere else. The wide range of mouth shapes and dental features open up a diverse package of feeding strategies. Some eat insects; some eat other fish; others eat mollusks, fish larvae, plants, or even the scales of other fish.

It is postulated that all these species are descendants of a single species, *Haplochromis bloyeti,* found in rivers feeding into Lake Victoria. There are several species in Lake Victoria that bear a considerable resemblance to this river species. Upon entering the newly forming lake, the unspecialized cichlids found fewer predators or competitors as well as a still water habitat as opposed to the flowing water of the river. This would set the stage for an adaptive radiation as outlined earlier. Thus it would seem that all of the punctuational criteria are met. There are many rapid speciation events leading to significant morphological changes, all in less that 1 million years. Concomitant with this is the persistence of the ancestral form that evolved only slightly by microevolutionary processes.

The island of Hawaii, the youngest of the Hawaiian chain, is estimated to be 750,000 years old also. Several species of honeycreepers are endemic to this island. As a whole, the honeycreepers of the Hawaiian chain have experienced wide diversification of bill size and shape that is even more dramatic than that found in Darwin's finches. This, of course, has led to a variety of eating habits.

These examples, collectively, are not used by Stanley to establish that the model of punctuated equilibrium is a certainty. Rather, they are merely meant to establish that punctuationalists are not asking the impossible. It is important to note again that rapid speciation was acknowledged by Neo-Darwinians; it simply was not a major component of the synthesis. But because macroevolution would be decoupled from microevolution as a result of species selection,

126

The Natural
Limits to
Biological
Change

Chromosomes,
Regulation, and
Bifurcations

many Neo-Darwinians rejected it. Speciation was always important, but never primary.

The last piece of the punctuationalist puzzle involves the genetic mechanisms of rapid morphologic change. More is involved here than simply rapid speciation. There must be real evolutionary change. There are essentially three areas of saltational effects currently being investigated: chromosomal alterations, regulatory gene mutations, and developmental bifurcations caused by the accumulation of minor mutations. However, as Gould is careful to point out, punctuated equilibrium is a descriptive theory of large-scale patterns over geological time, not a theory of genetic process. Rather, these saltational processes are part of what he calls punctuational thinking, which "focuses upon the stability of structure, the difficulty of its transformation, and the idea of change as a rapid transition between stable states."[104]

Chromosomal rearrangements have been the primary focus as the source of reproductive isolating mechanisms for the rapid development of speciation in small, isolated populations.[105] In sufficiently small populations, chromosomal rearrangements may become fixed in a handful of generations. The primary experimenters and theorists in chromosomal speciation are Bush,[106] Carson,[107] White,[108] and Wilson.[109]. In chromosomal speciation of this type, it is crucial to note that the formation of the reproductive isolating mechanism is a stochastic event that may have no adaptive value at all. However, although the speciation event itself may be nonadaptive, it may have peculiar results that affect the adaptedness of the organism due to new gene arrangements caused by the chromosomal alterations. A correlation has been shown to exist between rates of karyotypic diversity and anatomical diversity in certain vertebrate taxa.[110] Essentially all that is proposed is that since the arrangement of chromosomal material has been altered, certain gene complexes have been broken and new ones have been formed. This may sufficiently revise the regulatory program so that the patterns of expression of structural genes is changed.

This brings us to the next category of macromutation: regulatory genes. Regulatory genes function as activators or inhibitors of other genes. Regulatory mutations may bring about changes in the expression of genes coding for particular proteins. This can be accomplished with no effect on the amino acid sequence of the structural gene involved. Two types of regulatory mutations are recognized. First, mutations may occur in the regulatory genes themselves, such as the genes that code for the repressor or even the promoter or operator region. Such mutations can effect the quantity and timing of specific protein production. Second, the arrangement of genes on a chromosome may be altered, bringing together new gene combinations that may revise patterns of gene expression.[111]

Stebbins outlines five ways in which regulatory mutations may be significant in evolution: (1) the changes in the activation and inhibition of gene expression, (2) relative activity of different hormones and other growth substances, (3) alterations in the permeability of membranes, (4) alterations in cell shape, (5) and alterations in the frequency of mitosis and cellular proliferation.[112]

The potential for regulatory genes to effect evolution has long been noted by major supporters of Neo-Darwinism.[113] The major theorists were Britten and Davidson,[114] who noted that the majority of macromolecules in living organisms were similar in nature and that the major differences between organisms may be the result of revisions in regulation. Their model was then applied to the rapid appearance of major animal groups in the fossil record by Valentine and Campbell.[115] The substance of their proposal is that "animals responded to major environmental opportunities or challenges by enlarging or re-patterning the regulatory portions of their genomes, rather than by simply employing novel mutations or gene combinations within their structural gene complements."[116] By comparing morphologic divergence and structural gene evolution between frogs and mammals and between humans and chimpanzees,[117] Wilson and his colleagues have concluded that morphologic changes and structural gene divergence are independent. The change in morphology is then attributed to regulatory mutations.

As an example of the combination of rapid speciation and regulatory mutations bringing about sudden divergence, Stanley offers the giant panda. It was long a mystery whether the panda belonged in the family of raccoons or bears, but it is now aligned rather firmly with the bears. Stanley believes the panda is so distinct that it should be put into a separate monogeneric family. He proposes that the panda evolved by a quantum speciation event by way of a small population in a local bamboo forest. The marked morphologic divergence was accomplished by only a few genetic changes in regulatory genes.[118]

A final form of macromutation that has surfaced recently is a developmental bifurcation.

> When enough mutations (or a large one, depending on the nature of the genetic control system involved) accumulate to bring the developmental system near to a bifurcation boundary, then one can expect large changes in phenotype to accompany relatively minor genetic alterations.[119]

But first, what is bifurcation? A child's clicking toy frog or cricket clearly illustrates the principle. This is a simple device primarily composed of a piece of metal that is bent to the point that it buckles and emits a click. Releasing the pressure, unbuckling the metal, produces another click. The principle, then, is simple. By varying something smoothly and gradually, over time, a stress point may be reached where something dramatic may happen.

Oster and Alberch apply this concept to embryological development. Using skin structures as their model, they point out how evagination and invagination of the dermal matrix results in either scales and feathers or glands and hair respectively. In other words, small genetic mutations may accumulate to a point where a new developmental pathway may emerge, resulting in drastic morphological changes. The difference between evagination and invagination is small, but the phenotypic effect is quite noticeable.

Relationships between developmental constraints and morphologic divergence have been receiving increasing attention.[120] Many feel that it is an important factor that has been ignored for too long. Although Oster and Alberch believe that fixation of these new phenotypes in populations would occur in orthodox Neo-Darwinian terms, it is clear that any model for the rapid appearance of new morphological characters would be welcomed by punctuationalists.

It is indeed interesting to note that the punctuationalists are dusting off their copies of Goldschmidt's *The Material Basis of Evolution*.[121] Although not espousing his ideas of "hopeful monsters" to the letter, they do claim the essential essence of macromutation in common with him ("hopeful monster" is a phrase ivented by Goldschmidt to refer to a genetic monstrosity produced by a single mutation that might permit the occupation of a new environmental niche, thereby creating a new type in only one step). This brief excursion into contemporary concepts of macromutation would make Goldschmidt quite happy. Although not everyone's theory of macromutation is welcomed or even expounded by proponents of punctuated equilibrium, we can be certain that all legitimate theories will be received with enthusiasm.

Punctuated equilibrium has now captured wide attention. Gould's statement summarizes briefly the claims of the theory of punctuated equilibrium; it holds

> that evolution is concentrated in events of speciation and that successful speciation is an infrequent event punctuating the stasis of large populations that do not alter in fundamental ways during the millions of years that they endure.[122]

Accompanying these speciation events might be any one of a number of macromutation processes that would accelerate the rate of morphologic divergence. In order for the speciation event to be sufficiently rapid and for the macromutation to have a chance to become fixed, these processes undoubtedly have occurred in small isolated populations. This explains the lack of intermediates between species in the fossil record.

As proponents of any new theory that purports to replace an established orthodoxy will find, punctuationalists have not been unanimously received as brilliant new theorists. There has been opposition and criticism. The next chapter is devoted to a multifaceted critique of the basic tenets of punctuated equilibrium and its macromutational hangers-on.

chapter seven

Obstacles to Punctuated Equilibrium

Punctuated equilibrium has difficulties of its own because of the lack of testability of it major themes and the absence of reasonable genetic mechanisms.

Since punctuated equilibrium has emerged as a major competing theory of evolutionary change, it has aroused a storm of criticism. The tenor of these criticisms has ranged from impassionate claims of "nothing new" to rather vehement charges of "straw men" in regard to the unfavorable characterizations of Neo-Darwinism. But the debate is still new, and almost all participants are making claims and counterclaims that sometimes seem purely semantic in nature. Therefore, we will attempt to summarize only some of the major criticisms that have surfaced, as well as some major difficulties that are not often mentioned.

131

132

The Natural
Limits to
Biological
Change

Time, Sudden
Appearance,
and Stasis

Perhaps foremost among the arguments from Neo-Darwinists is that the paleontological observation of sudden appearance is purely an artifact of the difference of time frames studied by population geneticists and paleontologists. Thus they claim that there is no need for a major shift in emphasis from the modern synthesis:

> Instantaneous events in the paleontological scale, as in the transition between different geological strata, may involve thousands, at times many thousands, of years. In the microevolutionary scale of the population biologist, a thousand years is a long time, not an instant.[123]

Concomitant with this objection is the view that none of the claims of the punctuationalists are really very new and that the various mechanisms for rapid morphological change are an integral part of Neo-Darwinism. Some go so far as to say that in its general form Neo-Darwinism has not changed nor been seriously challenged since 1959, the centennial of the publication of *The Origin of Species*.[124]

Another area of comment on the paleontological observations of punctuated equilibrium involves the phenomena of prolonged stasis. This, too, has come under question. The first clarification that is needed is the qualitative difference between the biological species and a paleontological species. The biological species involves testable criteria of reproductive isolation, the existence of a definable gene pool that is isolated from all other similar gene pools. The species of paleontology is a morphologic species and has little to do with gene pools. Therefore, if a speciation event has occurred that does not yield a change in morphology, the paleontologist cannot recognize it and therefore assumes that no speciation has occurred. The claim that rapid morphological change in the fossil record is usually associated with speciation becomes tautological.

In answer to the question, "Can rapid morphological change occur in the absence of speciation?" we reply, "in the fossil record, rapid morphological change *is* speciation."[125]

In other words, if a new species is recognizable in the fossil record only by morphologic changes, then of course a paleontologist associates speciation with morphologic divergence. Adding to the confusion of definitions are the known cases of morphologically similar species that differ genetically (e.g., sea cucumbers, fruit flies, cotton rats, and seastars) and morphologically diverse species that are virtually identical genetically (e.g., pocket gophers, pupfish, minnows, and cichlid fishes).[126] These distinctions are not discernible in the fossil record.

A corollary to the above difficulty is the fact that only an organism's most resistant and numerous parts are preserved in the fossil record. Paleontologists are able to investigate only a small part of an organism's biological composition.[127] Schopf summarizes that skeletal changes observable in the fossil record account for only 10 percent of the possible avenues in which an organism may change.[128] Changes in internal organs, physiology, and behavior are not distinguishable to the paleontologist. The fossil record is, therefore, limited in the kind of evolutionary information it can provide. Consequently, the organism may be changing in profound ways, yet the paleontologist is unable to detect it. Stasis could, therefore, be 90 percent illusory.

Finally, in regard to stasis, Neo-Darwinists believe that stasis is best explained as a phenomenon "resulting primarily from stabilizing selection toward an intermediate optimum phenotype."[129] This is in contrast to the punctuational view that stasis is due primarily to a lack of substantial gene flow through the large population (thereby decreasing the probability of spreading a favorable mutation) and the onset of new developmental constraints due to the genetic revolution at speciation. If stasis is due to stabilizing selection, it is of course more compatible with Neo-Darwinism. In fact, Stebbins goes so far as to say that stasis will be very common on the basis of the genetic gradualism of the population geneticist, for genetic gradualism appears as punctuated equilibria to the paleontologists.[130]

<h2>Peripheral Isolates and Genetic Revolutions</h2>

The crux of punctuated equilibrium to the biologist revolves around the various models of rapid speciation accompanied by morphologic divergence. This is most unfortunate, however, because this is one area of population biology that remains a very large mystery. In 1974 R. C. Lewontin made the often quoted statement that "we know virtually nothing about the genetic changes that occur in species formation."[131] Earlier in the same paragraph, Lewontin notes with irony that evolutionary genetics had made no direct impact on the theory of species formation.[132] Indeed, Darwin himself did not address the problem of the origin of species, although it is the primary title of his book.[133] Recently, Guy Bush has echoed the above frustration: "Although the importance of speciation is clear and convincing, the processes involved are, for the most part, unknown."[134] Bush also declares that it is the uniqueness of each speciation event that "makes it difficult if not impossible" to produce a comprehensive theory of speciation that is applicable in the real world. One is left with highly subjective post hoc reconstructions of already-completed speciation events.[135]

This significant problem has not been mentioned frequently as a criticism of punctuational schemes based on speciation. With a heavy emphasis on species selection and rapid species formation, concurrent with morphologic divergence, punctuational models appear to find themselves in a precarious position. As stated earlier, Stanley admits to having very few examples.[136] This raises the question of whether punctuated equilibrium is empirically testable in its present formulations. The speciation process, on which it depends in its biological aspects, is nearly impossible to observe; it is, therefore, extremely difficult to devise meaningful tests. This crucial weakness, along with the tautological nature of species identification and speciation in the fossil record, makes any punctuational model difficult to present as a scientific theory. The claims that punctuated equilibrium is only a paleontological description become rather hollow in light of the many biological processes offered by punctuationalists to support their contentions. In fact they only serve to emphasize the speculative nature of punctuationalism.

What little evidence is available seems to suggest that genetic revolutions accompanying speciation from small peripheral isolates are not the usual case. The concept of genetic revolution is based on the idea that the gene pool of a species is a tightly knit, coadapted unit.[137] Upon speciation by founder effect, this cohesive unit is broken down, and the genome is allowed to rearrange itself. In time, a new coadapted complex evolves. In the Hawaiian *Drosophila*, in which such revolutions would be expected to be found, they have not.[138] Furthermore, the enzymatic proteins studied by electrophoresis have been suspected of being relatively insensitive markers of the speciation process.[139] Therefore theories about genetic changes during speciation based on this data are questionable. This is verified in part by the observation made earlier of the existence of genetically dissimilar sibling species and distinct morphologic species that are similar genetically.

A recent study undertaken to indirectly test the amount of morphologic divergence accompanying speciation was done by Douglas and

Avise.[140] Morphologic divergence among a species-rich genus of minnows (*Notropis*) and a species-poor genus of sunfish (*Lepomis*) were compared. Using multivariate morphometric methods, Douglas and Avise showed that despite prolific speciation, the species of minnows were no more morphologically differentiated among themselves than were the few species of sunfish. This is another result inconsistent with punctuated equilibrium. If morphologic divergence is to accompany speciation, these species-rich groups like *Notropis* should show greater morphologic divergence than species-poor groups.

A final note in regard to speciation by peripheral isolates involves their survivability. Mayr reflects that 95-99 percent of all peripheral isolates will go extinct before ever completing the speciation process.[141] Survivorship is obviously not very good. Though there are a number of random factors that could be involved, the key factor is that the population is particularly vulnerable in the midst of the genetic revolution. This period of revolution would be an extremely vulnerable time. Even once the speciation process is complete, survivorship would not be expected to be very great, since the population would probably be adapted to a rather unique and limited environment that increases the probability of extinction by stochastic events. The percentages are reduced even further. Consequently, not only is the discovery of an intermediate peripheral species in the fossil record unlikely, but the overall prospects of survivability of the peripheral isolate and the reality of a genetic revolution would seem vanishingly small.

Chromosomes and Macromutations

Chromosomal rearrangement and macromutation both have potential as a means of rapid speciation and rapid morphologic change with punctuational schemes. However, the prospects for determining in detail what role these processes of chromosomal rearrangement may actually play in speciation are extremely limited. Macromutation is also a phenomenon in which the researchers are simply going to have to be rather fortunate in their timing to truly document it as having occurred and as being of adaptive significance to the population. Finally, in regard to the fossil record, both phenomena are undetectable, chromosomal rearrangements for obvious reasons, and macromutations will be indistinguishable from changes that may have occurred gradually yet may not have been preserved in fossil form. Once again, we wish to emphasize the highly speculative nature of these proposed punctuational mechanisms.

That chromosomal rearrangement has played some role in speciation is little disputed. One need only look at the often subtle changes in karyotype between closely related species and populations.[142] North American pocket gophers of both the *Geomys* and *Thomomys* groups show extensive karyotypic variation between populations of the same species,[143] the only critical morphologic difference within the genera being size. One of the authors of this book was involved in a study that indicated that perhaps the karyotopically divergent populations within species exhibited sufficient restriction in gene flow to be categorized as functionally distinct species.[144] If this were found to be true in other intraspecies relationships, the significance of karyotypic divergence in this taxon would be greatly increased. However, as Bush has pointed-ed out, the role of chromosomal rearrangement in speciation depends heavily on various organismal and ecological characteristics.[145] Thus chromosomal rearrangements are not ubiquitous in regard to speciation events. The Hawaiian *Drosophila* again show very clearly that many speciation events can take place with virtually no chromosomal restructuring at all.[146] Also, others have pointed out that related, yet karyotypically divergent, species usually differ by many minor, more easily fixed rearrangements than one major chromosomal change.[147] Besides, it is extremely difficult to determine whether the chromosomal rearrangement was instrumental in the speciation event or whether it took place after the fact. This would indicate a rather stepwise process toward reproductive isolation. It would appear that major chromosomal adjustments are not easily accomplished and certainly not in a single event.

The overriding consensus of geneticists and population biologists, by far, is that the feasibility of macromutations being significant in evolution is practically nil.[148] Primarily, their argument stems from a principle developed early by Fisher that the likelihood that a particular mutation will become fixed in a population is inversely proportional to its effect on the phenotype.[149] Although this is an argument straight from Neo-Darwinian orthodoxy, it does have a great deal of merit. A mutation of small effect is not very likely to disrupt an organism's balance with its environment, whereas a mutation producing a drastic phenotypic change stands the possibility of greatly altering the organism's chances of survival. Of concern is not whether such mutations occur. They do. What matters is, first, whether they are capable of surviving and, second, whether the mutation can be assimilated by the whole population, even a small founder population. Apparently there is little if any evidence to indicate that either of these situations is possible.

Regulation and Development

The role of regulatory mutations in evolution has become popular even outside the camp of the punctuationalists. However, here again is an area for which there is little, or at least questionable, factual evidence to back up the claims. Only a few copies of a regulatory protein may be produced, making biochemical assays difficult, to say the least. Then, there are those regulatory areas of the operon that produce no protein product at all. To study the effects of regulatory mutations directly will prove to be no easy or small task. What little evidence there is to support the role of regulatory mutations comes largely from inference. In the study by King and Wilson comparing genic divergence between humans and chimps, these two widely divergent morphological species were found to be astoundingly similar in the structural loci. Four proteins were compared immunologically, twelve by amino acid sequence, and forty-three by electrophoresis. All these showed remarkable similarities. The only test that showed a sizeable difference was nucleic acid hybridization. Therefore, since there is little difference among structural genes, the differences in DNA sequence may be due to differences in regulatory genes. On the other hand, the fifty or so proteins may have been inappropriate to establish genetic distance in this case. Morphological differences may indeed be controlled by structural loci, just not the ones studied. Fifty proteins are, after all, a small percentage of the ten thousand to one hundred thousand suspected genes in humans. Many proteins should be expected to be similar for all mammals. These particular proteins were chosen for their ease of study and for the useful comparison of the results of their study to results from the study of other taxonomic groups, not for their significance in differentiating between humans and chimps. Regulatory mutations are therefore one possibility, but not necessarily the most compelling.

Stanley's use of the giant panda as a possible case of a few mutations (regulatory) accompanied by rapid speciation is highly conjectural. The study on which he bases his theory on has been sharply criticized,[150] and it has been pointed out that the whole scenario has been based on practically no genetic information.[151] The giant panda is not so easily explained. The fact that so

**Regulatory
Mutations**

much of its behavior remains enigmatic cautions against speculating on its origin with so little factual evidence.

The role of regulatory mutations should be demonstrated during embryonic and postnatal development. The claim that large phenotypic changes can be achieved through small mutations affecting the timing of developmental events is hardly questioned.[152] The real question is whether or not these mutations in development can have anything to do with rapid evolutionary change. The development from a newly formed zygote to an independently functioning multicellular organism is a tightly organized and precisely regulated process. Small changes can indeed have profound phenotypic effects. This is especially true if the change occurs early in development. Unfortunately, virtually all discussions of developmental gene mutations are discussions of lethal and near-lethal consequences. Mutations of genes intricately involved in development are what produce hopeless, not hopeful monsters. One reason for this is the common phenomenon of pleiotropy. Many genes, if not most, exhibit multiple phenotypic effects. In other words, a mutation in one gene does not limit its effect to only one phenotypic character. There may be many, seemingly unrelated, phenotypic alterations due to the mutation of a single gene. This, undoubtedly, is true at the level of regulatory genes and genes heavily involved in development, severely straining belief in survivability of such mutants.

Actually, the tight coordination of developmental pathways may serve both to minimize the effect of mutational change and to define the limits of biological change. A minor change in one gene will usually have little or no effect on the basic pattern. Such a developmental pattern is said to be buffered or canalized. This protection effectively resists change brought on by genetic mutations. The primary model of Oster and Alberch in their theory of developmental bifurcation is the development of scales, feathers, hair, and skin glands from epithelial tissue. However, there is no information supplied as to what the genetic factors are that control these phenomena. In addition, it must be remembered that here also the reality of bifurcations is not at issue. The overriding problem is whether such bifurcations can be expected to produce organisms capable of adapting to a new environment or of being more effective in the old. Also, the accounting of novel structures is necessary. Examples of salamanders that have reverted back to nymphlike structures in the adult phase is of little help.[153] Such examples may explain diversification within a group, but they do not contribute to the search to explain how birds came to fly and how whales came to live in the sea.

The overall factor that has come up again and again is that mutation remains the ultimate source of all genetic variation in any evolutionary model. Being unsatisfied with the prospects of accumulating small point mutations, many are turning to macromutations to explain the origins of evolutionary novelties. Goldschmidt's hopeful monsters have indeed returned. However, though macromutations of many varieties produce drastic changes, the vast majority will be incapable of survival, let alone show the marks of increasing complexity. If structural gene mutations are inadequate because of their inability to produce significant enough changes, then regulatory and developmental mutations appear even less useful because of the greater likelihood of nonadaptive or even destructive consequences. This is primarily the result of the crucial role these genes play in integrating the many life processes of an organism. Much remains to be discovered in this relatively new field of research. But one thing seems certain: at present, the thesis that mutations, whether great or small, are capable of producing limitless biological change is more an article of faith than fact.

Although the punctuationalists' major argument stems from their interpretation of the paleontological data, Stanley in particular goes to some length to document examples of living forms that exhibit rapid speciation concurrent with significant morphologic divergence. This is done purportedly to show that punctuationalists are not asking the impossible. Though Stanley offers many examples of rapid speciation, as we detailed in chapter 6, he readily admits that most offer no help in the primary criterion of rapid morphological change. Those few he does put forward as showing marked morphological divergence are also highly debatable.

The desert pupfish of Death Valley is a curious example. Here is a group of similarly adapted fish, very similar to one another except in size, on the precipice of extinction because of the rarity and fragility of their unique environment. This hardly seems the place to document punctuationalists' claims. The desert pupfish rather seems to exhibit all that is wrong with punctuated equilibria. When the desert environment was initiated, strong selective pressures fragmented the initial population. Those small pockets that survived represented the far limits of heat and salinity tolerance. Upon specialization, their position is a precarious one, with all indicators pointing to extinction. With genetic variability split by the speciation process and dwindled by severe selection, these fish have little hope of adjusting to even minor perturbations in their environment. It is most likely that the desert pupfish of Death Valley are at a dead end. A rapidly speciated and narrowly adapted group, formed out of desperation, is headed for extinction.

The cichlid fish of Lake Victoria may, at first,

seem to offer a more convincing case. The time span for the development of these fish is far longer than for all the other examples, but then again, the number of species is also far greater. The watershed issue again comes down to the degree of morphologic divergence. The major disruption is the diversification of mouth parts, a diversification that leads to varied feeding habits. But has significant evolutionary change really taken place? Stanley points out that recent studies have shown that cichlid fishes have a complex jaw structure. The rear portion is decoupled from the front. The activities of gathering and processing food are separate from each other. This allows for a great deal of diversification of feeding habits[154] The level of divergence seen in mouth parts and feeding habits of the Lake Victoria cichlids is to be expected on the basis of the overall morphology of the mouth as a whole. This could simply be due to novel combinations of genes present in the initial species. This remains an interesting example of adaptive radiation. What remains open to question is whether true evolutionary novelties have been produced. New mutations may have been totally unnecessary.

The example of the Hawaiian honeycreepers may prove more cooperative. Although the degree of divergence in beak size and shape is dramatic, one questions how novel these adaptations actually are. This is certainly not on the scale of change from a tiny shrewlike mammal to the whale. One must ask why these birds radiated and others did not. The same question could be asked concerning Darwin's finches. These are not the only birds on the islands. That they may have arrived first is one possible explanation. By being among the first arrivals, they could have radiated to fill the various niches before other species arrived. On the other hand, the initial population may have simply contained a large storehouse of genetic variability and phenotypic plasticity. Stanley points out that the basic whale design had to evolve in less than 12 million years. In the estimated 5-million-year existence of the oldest Hawaiian island, nowhere near this kind of divergence has taken place in the honeycreepers. They are all still easily relatable within the same low-level taxon. Another note to make is that the honeycreepers are also fighting extinction. To say that this is largely due to man's influence is to miss the point. These rapidly radiated species have been unable to cope with environmental fluctuations. This is as we would expect with species arising by fragmentation of a large gene pool. By speciating rapidly, probably by founder events, their storehouse of genetic variability is drastically reduced, limiting the extent of adaptive response following establishment.

Another case involving the rapid origin of significant morphologic change that is not cited by Stanley (probably because it does not involve speciation) is the appearance of extra limbs in frogs remaining in their natural environment.[155] This has been widely observed in isolated populations of frogs in many parts of the world. In these populations the frequency of individuals with supernumerary limbs can be as high as 30-33 percent. Most often these legs are useless to the frog. There have been rare reports of limited degrees of functionality. The phenomenon is thought to result from the influence of some environmental factor such as a virus to which some frogs respond and some do not. Van Valen used this example as a springboard for formulating a theory for the discontinuous origin of higher taxa.

However, it is difficult to envision how this observation has anything to do with evolution. First, it has been determined that the formation of extra legs is environmentally induced. Some frogs respond; some do not. Second, in order for this to spread completely through the population, it would require both a genetic component and a distinct advantage to having an extra set of legs. At this point, the only possible heritable element is the susceptibility to the environmental component. But this is not known for sure. Third, at some point, it would seem preferable that the extra legs would be expressed without the environmental component present. This is a process difficult to imagine. The adaptive significance is perhaps the highest hurdle to overcome. Of what possible use is an extra pair of legs to a frog? The only suggestion offered by Van Valen involves the potential need for additional thrust in a more viscous environment. No potential real-world instance is provided, however. Van Valen even suggests that it is not really necessary to propose a real environment to which the frogs are preadapted. It is only necessary that other abnormalities occasionally be so preadapted. This appears to be begging the question. In almost all instances these extra limbs are nonfunctional. Those that are functional do not seem to affect the adaptiveness of the frog.

The persistence of this nonadaptive phenomenon could perhaps be explained by two factors. First, the ability of amphibians to regenerate limbs is well known. Therefore it is not really surprising to see extra limbs develop as a result of some environmental miscue interfering with developmental signals. Second, it is apparent that the nonfunctional and the functional legs are not ill-adaptive; otherwise the supposed genetic susceptibility would be selected against, eventually making the phenomenon an extreme rarity. Alternatively, no studies have been undertaken to determine the extent of the genetic susceptibility. It may be so slight that selection would be rendered virtually ineffective.

Punctuated Equilibrium in Perspective

One issue that the various schemes of macromutation fail to address involved the addition of new genetic information. The difficulties of gene duplication have already been discussed in chapter 5. The punctuationalists offer no new alternatives or refinements in this regard. We emphasize again that this is crucial. No theory of evolutionary change is complete without some workable mechanism for generating new genetic information.

The same criticism can be offered in terms of mutations in general. There appears to be a general lack of appreciation as to what a mutation is and what its effects on the organism may be. Discussions of regulatory and developmental mutations are carried out with no regard to the overwhelmingly destructive effect such mutations will likely produce compared to mutations of structural genes. There is also a misleading tendency to ignore just how little we know of regulatory processes and how much of a mystery embryonic development continues to be.

This brings up the recurrent theme of criticism of punctuated equilibrium: its reliance on processes and patterns that are either presently little understood or virtually unknowable. Regulatory mutations certainly fall into the first category. There is also the species identification problem in the fossil record, the tautology of associating morphologic change with speciation in the fossil record, the ephemeral nature of peripheral isolates, and the uniqueness in history of speciation events.

One is left with a similar problem as that of the Neo-Darwinists who attempt to extrapolate from observable microevolutionary processes to unobservable macroevolution. How do we know it really occurs? A beginning student of evolution finds himself in a serious dilemma. In searching for solid evidence of evolutionary change, he first turns to the biologist. The biologist reports, however, that dramatic evolutionary change is such a slow process in our time frame that such documentation is virtually impossible. He suggests that the student visit a paleontologist. All the paleontologist shows, however, are systematic gaps in the fossil record. Geologically speaking, evolutionary novelties appear so quickly that there is little hope of documenting such changes in the fossil record. Although the example is a slight exaggeration, the point is clear. Evolution either occurs too fast or too slow for real observation.

Our frustration is compounded by a second disagreement between disciplines. The English zoologist Mark Ridley maintains that the evidence for evolution does not depend at all on the fossil record. The real evidence comes from microevolution, biogeography, and taxonomy.[156] In direct contradiction is the view of the American paleontologist Stephen Stanley, who states that without the fossil record, evolution would appear as little more than an outrageous hypothesis. The fossil record is necessary to draw conclusions about large-scale change. Evidence from biology only leads to inferences about evolution.[157] The real evidence for evolution, whatever it is, seems to depend on who is doing the talking. We can't help but recall the story of the emperor's new clothes. The proponents of each theory cite the evidence as supporting their own particular theory. But when we try to examine the details, the reply is either "Our discipline really doesn't have the tools to know that" or "The other discipline is also unable to document that." On the other hand, as we have pointed out, there are also alternate explanations that are frequently left out.

It is common to hear of evolution as both fact and theory. The fact is that it occurred; the theory concerns how it occurred. This is in juxtaposition to other areas of science. Take continental drift, for example. Here was a theory long ridiculed primarily because nobody could figure out just how continents were supposed to move. But with the recognition of plate tectonics in the late 1960s, suddenly there seemed to be a mechanism, and it revolutionized geology, even though it remains unclear as to why the plates are as they are. Suddenly, all the circumstantial evidence had meaning: the fit of the continents, correlation of fossil material, etc. Continental drift was no longer a laughing stock when it was without a mechanism.[158] Evolution is supposed to be a fact, yet like continental drift earlier, it consists mainly of circumstantial evidence, with no documented mechanism. One gets the feeling that a double standard is in operation.

In regard to punctuated equilibrium, let us close with a quote from Stephen Stanley:

> Any claim that natural selection operated with great effect exactly where it was least likely to be documented—in small, localized, transitory populations—would have seemed to render Darwin's new theory untestable against special creation, and perhaps almost preposterous as a scientific proposition.[159]

Although Stanley's remark was offered in a historical context, we wonder why punctuated equilibrium is any more testable today than it would have been in Darwin's time, particularly in view of the preceding discussion. Perhaps it is time to consider more seriously and openly a long-standing alternative in light of modern biology.

chapter eight

Another Alternative

The available data of biology indicates that in contrast to evolutionary theories there is sufficient evidence to suggest that biological change has limits. More specifically, if the rationality of a theistic world view is entertained, a new theory of biological origin and change emerges.

As we have tried to point out, the two major competing models of evolutionary change are far from being considered accepted facts of nature. Both suffer from serious problems from which, some say, they may not be able to recover. This, of course, must stand up to the ultimate test of time, buffered by further discovery, particularly of the genetic machinery. However, if one sits back and views the evidence as a whole, a totally different perspective arises as a possibility. Neo-Darwinism and punctuated equilibrium both subsist on the view that all of life is   149 biologically related. By simply combining processes of mutation, natural selection, recombination, speciation, etc., all of life's manifold variety, from the simplest prokaryotic bacterium to man, can be accounted for—nature subject to virtually limitless biological change. But is this really what the data are telling us?

A Quick Review

First, virtually all taxonomic levels, even species, appear abruptly in the fossil record. This, it will be remembered, is one of the sharper criticisms of Neo-Darwinism, and one of the two cornerstones of punctuated equilibrium. It is relevant not only that the various levels of taxa appear abruptly but also that alongside the higher taxonomic levels there are unique adaptations. This is the key. Unique and highly specialized adaptations usually, if not always, appear fully formed in the fossil record. Two clear examples are the origin of swimming adaptations and the origin of flight. Anderson and Coffin document these two phenomena carefully.[160] Numerous examples of aquatic reptiles and mammals are shown fully adapted to an aquatic existence. Flight is present in four distinct groups: in the phylum Arthropoda—the insects; and in the phylum Chordata—the birds, reptiles, and mammals. Notable contrasts are the whales and the bats: both these mammals supposedly evolved from the same early mammalian ancestor in less than 10 million years.[161] Our imaginations are stretched, no matter which evolutionary mechanism is considered. How could two totally different organisms evolve from another totally different organism in such a short time or any amount of time?

Second, there is the steady maintenance of the basic body plan of the organism through time. This is essentially the theme of stasis in punctuated equilibrium. One need only think of the living fossils from paleontology and of bacteria and the *Drosophila* fruit flies from genetics. The basic body plan, or *Bauplan* as some have called it, does not change whether analyzed through time in the fossil record or through mutations in the laboratory. This conclusion is reinforced by animal and plant breeders through artificial selection. There is much variation, but it can be manipulated only to a limit. And, as stated in chapter 5, a mutation may extend the limit, but it cannot break the constraints.

Third, we find that the observed adaptational change is predominantly brought about through recombination and selection on variation already present in the population. Mutations, when they do play a role, produce a defective organism that survives and thrives only in an unusual and unique environment. The chances of outcompeting other, similar organisms (e.g., mutated bacteria versus wild-type bacteria in a more natural environment) are minute.

Fourth, and this is more of a corollary of the previous two, we see the apparent inability of mutations to truly contribute to the origin of new structures. The theory of gene duplication in its present form is unsuitable to account for the origin of new genetic information that is a must for any theory of evolutionary mechanism. If mutation is to be the ultimate source of all genetic variation, then its observable track record in producing truly useful variation does not exactly place it as the gambler's favorite.

Fifth, we observe the amazing complexity and integration of the genetic machinery in each and every living cell. First came Mendelian genetics, followed by the elucidation of the structure of DNA and the genetic code. The arrangement of genes on a chromosome is another case in point. It was formerly thought that the genes were lined up on the chromosome like beads on a string, one right after another. Then it was discovered that often the genes were separated by long segments of nontranscribed DNA, these sequences often being repetitive. The discovery of regulatory genes and the operon of prokaryotes complicated matters even further. But all seemed to be overshadowed by the discovery of introns, intervening lengths of nontranslated sequences within the gene itself. Some exons or translated portions are now also found to code for discrete functional units in the final protein product.[162] What we do know of the genetic machinery is impressive; what we have yet to learn staggers the imagination. One's curiosity is aroused as to how the machinery itself came about, let alone how mutation, selection, and speciation could ever hope to improve or change the machinery in any substantial way.

The sixth and final element involves the big picture. Ecosystems themselves are a marvelous balance of complexity and integration. Interdependence of diverse organisms is imperative. As yet we have only the remotest of understanding about the workings of the complete ecosystem. One can devise schemes of energy flow or biomass flow through an ecosystem as complicated as any biochemical pathway or genetic regulatory scheme. At the center of all this is the wondrous fit of an organism to its environment. It was the unique adaptedness of an organism to its own peculiar environment that, in the time before Darwin, was the chief evidence of a Supreme Designer. This went along with the near-optimal construction of structures and organs themselves.[163] Our questions, therefore, are simple: Has the situation really changed all that drastically since then? Has Darwin's theory of natural selection really shown intelligent design in nature to be unreasonable? And in view of the failure of the evolutionary mechanism to be convincing, might biological change be a limited affair? Could the limited nature of biological change arise from the very nature of the genetic code itself, the unique set of structural and regulatory genes present in various groups of organisms, and the tight organization and coadapted nature of the entire genome? Our basic model proposes that, yes, there are limits to biological change and that these limits are set by the structure and function of the genetic machinery.

Design in Nature

Design in nature has been a topic of debate for centuries. We have no intention of trying to argue conclusively on a philosophic basis that intelligent design disproves any evolutionary theory. Rather, we hope to show briefly by example that the need for intelligence in bringing about the various designs in nature is by no means a preposterous proposition. However, for an interesting recent argument supporting intelligent design in the universe, see James Horigan's *Chance or Design*.[163] We maintain that there are two recognizable and yet intuitively distinct forms of design: that which is easily explainable in terms of physical properties and processes and that which is most readily explained by intelligent ordering.

Some contrasting examples will make the point. When a water droplet freezes around a dust particle, an elaborately designed snowflake results, simply because of physical properties. The same is true for the development of a crystal in a supersaturated solution. Just let the solution cool and then tap it. Order suddenly emerges as the crystal develops. On a larger scale, we can observe in the American West numerous examples of natural sculpture: chimneys, arches, bridges, human profiles, etc., all dutifully sketched from the rock by the forces of wind and rain. But one would not confuse such structures with Mt. Rushmore, the Golden Gate Bridge, or the World Trade Center. These human architectural artifacts show clearly the marks of intelligent design. The sharp geometrical forms and the finely sculpted lines are recognized intuitively as the works of human intelligence. The simple point is that intelligent design is discernibly different from natural design. In natural design, the apparent order is internally derived from the properties of the components; in creative design, the apparent order is externally imposed and confers new properties of organization not inherent in the components themselves.

Carl Sagan uses the same reasoning in describing the search for intelligent life on Mars.[164] Initially, the canals on Mars, as visible through telescopes, were thought to be possible evidences of intelligent life. Because of the patterns involved, it was considered possible that the canals were engineered by intelligent beings. However, on closer inspection, first by space probes passing close by and eventually by probes actually landing on Mars, the canals seemed to fade into nothingness; apparently, they were superficial artifacts and nothing more. However, as one approaches Earth, the conclusion is the opposite. From the distances of space, there is nothing about Earth's surface that betrays the presence of intelligence. Continents and seas have no observable patterns. But as one gets closer and sharpens the focus, geometry is revealed in the forms of farms, fields, highways, streets, and buildings. Suddenly the presence of intelligence becomes unmistakable.

With living things, the best of both perspectives is readily apparent. From a distance, biological organisms appear marvelously adapted to their environments indeed. They slither, crawl, walk, run, and fly, and some go nowhere at all. If we can be excused for speaking anthropomorphically, we can find beauty, humor, mystery, and drama in nature. Chapter 2 told only a few of the amazing stories of adaptation to be found that are truly exciting to the natural historian. But what happens on closer scrutiny? Does the apparent design fade into nothingness? On the contrary, it explodes into a maze of integrated complexity. Not only do geometrical patterns reemerge in microscopic form as in the feather of a bird or the scales of a butterfly wing, but at the molecular level, volumes are needed to express all that goes on. Inhibition, initiation, and feedback mechanisms produce a complex package of tightly regulated biochemical pathways. Orchestrating the whole menagerie is the genetic machinery—the beguilingly simple molecule DNA.

Discerning the Mark of Intelligence

The mark of intelligence is not exactly hard to discern. It will be helpful to recall that in chapter 5 we made note of the extensive use of language terms in describing DNA and the formation of protein. *Code, transcribe,* and *translate* are not just convenient terms but they accurately describe the process involved. If the DNA contains coded information, then some knowledge of human codes and their transmission would be of use. This brings us to the topic of information theory. Although there will be more on this later, suffice it to say at this time that information codes, of which DNA could be said to be one, require intelligent manipulation, not only to create the vocabulary but also to set up the rules of transmission. For our purposes, the analogy can be made as follows: Suppose we were walking down the beach or a river bank and, as might be expected, we observed ripples in the sand. Ripples are a perfect example of natural order easily derived from the physical properties of the sand and water and the movement of water in waves over the sand. But let's suppose that we decided to walk farther. Later we came across the words *JOHN LOVES MARY* written in the sand. This time we intuitively recognized the writing as intelligent manipulation of matter to formulate coded information.[165] This, then, is order externally imposed on the sand. The properties of sand and water alone are unable to produce this level of complexity. In DNA's language there are precise symbols or letters (nucleotides) grouped into words (codons). These words are also arranged in a particular sequence to form sentences (genes) or complete thoughts. The analogy can go on with paragraphs (operons), pages or chapters (chromosomes), and volumes (genomes).

By simply applying the same basic criteria to living systems, we find that the evidence of intelligent design is impressive, if not compelling. Since we are dealing with a historical question regarding the origin of the complex arrangements, it may be possible to construct a complicated mechanistic scenario to account for it apart from intelligence. This, however, would not prove that intelligence was not used, nor would it prove intelligence unnecessary, for any mechanistic theory will have a low probability. All that can be said is that it *may* have happened this way. Ultimately, the choice will be made (as we will elaborate later) based on personal preference, rather than on an inescapable conclusion forced by the facts.

We believe that the application of information theory to the field of genetics will yield a comprehensible theory of limited biological change. We will attempt to present a model of limited biological change constrained by the genetic machinery yet endowed with an impressive display of variability and hence adaptability. This is essentially the creationist's concept of "the created kind."

Various attempts have been made to suggest theories for delineating the boundaries of a kind. Marsh's theory of gametic fusion states that only organisms within a kind will produce a successful fusion of gametes with varying degrees of success following zygote formation.[166] But this approach is rather simplistic and offers no rationale for why this is so. Jones offers a working hypothesis that states that the basic organizational pattern (kinds) are encoded in membrane templates called cortomes. Cortical or cytoplasmic inheritance has been noted in protozoa and multicellular forms. In Metazoa this is found primarily during early development stages after fertilization.[167] The major means of recognizing the created kinds would be through behavioral patterns.[168] We see a number of problems here as well. First, the behavioral criterion will distinguish only the animal kinds; plants are not mentioned. Second, it is difficult to imagine how the membrane of a fertilized egg carries the major portion of the information for the basic organizational pattern that develops. That is the power and simplicity of DNA. The cytoplasm does seem to carry certain cues that turn genes on or off during development, but it is still DNA that carries the bulk of the information. While we appreciate and benefit from the effort of these authors, we intend to present a creationist model of limited biological change based on the genetic structure of the organism.

Genetics and Information Theory

In recent years several authors have explored the connection between the genetic code of DNA and information theory.[169] This is wholly reasonable because information theory concerns itself statistically with the essential characteristics of information and how that information is accurately transmitted or communicated. DNA is an informational code, so the connection is readily apparent. These studies, for the most part, have concerned themselves specifically with the origin of the genetic code. The overwhelming conclusion is that information does not and cannot arise spontaneously by mechanistic processes. Intelligence is a necessity in the origin of any informational code, including the genetic code, no matter how much time is given. The old analogy of a monkey sitting at a typewriter and, given enough time, producing the complete works of Shakespeare is foolish. On a computer, William Bennett set one trillion monkeys to typewriters, typing ten keys a second at random. We would have to wait a trillion times the estimated age of the universe before we would even see the sentence, "To be, or not to be: that is the question." It may not be theoretically impossible for a pot of water to freeze when placed on a lighted stove burner, but the *real* probability is so absurd that it is hardly worth talking about.[170] The same is true of monkeys typewriting Shakespeare.

More directly, though, our concern is with what happens after the code is in place. Similar to what happens in language, there are two fundamental principles involved in the expression of genetic information. First, there is a finite set of words that are the essentials of content. In organisms, this is comparable to structural genes. Second, the rules of grammar provide for the richness of expression using the finite set of words. In organisms, these rules or programs consist of the regulatory mechanisms. In human language, given a finite set of words and a single set of rules, the variety of expression goes on and on. In fact, in information theory, the more interesting question is not, "What does this message say?" Rather, it is "Why was this expression of a thought chosen over the countless others that could have done the job?" But the message essentially communicates the same information. So there is uniformity of meaningful content, yet a rich variety of expression of that same content. This concept could provide the basis for a model of extensive variability within the created kind based on one set of rules or regulatory mechanisms. When one adds allelic differences, the variety of expression becomes even more extensive. There are hundreds of species of fruit flies, but all are alternate expressions of the same basic fruit fly type or pattern (species alleles).

This could also provide the clue as to what mechanism preserves the integrity of a created kind. We have already noted the observation that what separates the higher categories are perhaps the regulatory mechanisms and not just structural genes. It is conceivable, therefore, that the different kinds are characterized by slightly different regulatory mechanisms, i.e., different programs. Certainly there are structural genes present in a mouse that are not to be found in a bacterium, but these structural gene differences may distinguish only major categories such as kingdom, phylum, or class. Structural gene differences within the class Mammalia may be very subtle, amounting to simple amino acid substitutions in the same protein. The distinguishing characters here are whole genes, structural information found in one organism but not present in another. However, with a different set of rules, a new organizational pattern is established that is also rich in the variety of possible expressions. The regulatory mechanisms would simultaneously both constrain the kinds within their organizational pattern and provide for variety. The created kinds, therefore, maintain their integrity in the face of a vast array of variation by virtue of their own unique set of regulatory mechanisms. Developmental biology, and here we agree with A. J. Jones, will prove to be the major proving ground for this approach.

On the other hand, one might say that if language is so powerful in its ability to produce variety and novelty, couldn't mutation and natural selection change the rules of regulatory mechanisms to produce biological novelty? The answer lies in the origin-of-life question. Informational codes are constructed of vocabulary and grammar. Both, of necessity, are produced only by intelligence. To argue that the genetic information in DNA originated initially as random nucleotide interactions seems analogous to claiming that the word processor, rather than the person operating it, actually authored a given book. Random changes in letter and word sequences ultimately can produce only gibberish. The same will result if one attempts to change the rules. Jeremy Campbell, in his book *Grammatical Man,* discusses this very possibility.[171] In his attempt to produce a unification of information theory and evolution, he makes some rather inexcusable leaps. The primary vehicle for the addition of new information would be gene duplication. Subsequent mutation to the now-redundant gene would produce a new gene. Since the new redundant gene would be free from selection pressures, it could mutate rather freely. After admitting that most of these sequences would degenerate into gibberish, mere genetic noise, he suddenly leaps to the following propositions:

> In some cases . . . mutations may result in a gene copy acquiring a meaning. A structural gene would code for a useful new protein. A gene that was part of a regulatory system would alter the timing of expression of the structural gene in a way that could be of benefit to the organism. These extra pages would then add sensible new information to the book.[172]

But this flies in the face of all that we know of the origin of informational codes. Words do not descend to gibberish by random changes only to ascend back to a new meaning by the same process. Campbell appears to abandon the foundation already laid. As biologists, we also wonder how the redundant gene slips out of the selective process only to slip back in once the new meaningful gene is complete. Was the copied gene no longer translated into protein while mutating? And if so, how was it included in the process again later on? We have the uncomfortable feeling that someone is simply waving a magic wand.

Campbell may have been harking back to the monkeys at the typewriters. Bennett went further in his simulations and began applying rules to the process based on an analysis of "Act Three" of *Hamlet*. As more rules were progressively added (more constraints), the sequences became more and more Shakespearean.[173] But where did the rules come from? They came from the mind of Shakespeare and were imposed on the computerized monkeys by the mind of Bennett. The initial rules did not arise by chance. Additional rules did not result from random modifications of the existing rules. To gain higher levels of complexity, new rules had to be added. If anything, the example supports our contention that the new information and programs that distinguish the kinds have their origin in the intelligence of their Creator. With structural genes, gene duplication produces only genes of similar function; nothing new emerges. Duplication of regulatory genes can only arrive at the same end. This, we believe, will be supported further by the advances and application of information theory to the field of genetics.

Admittedly, our proposal of regulatory mechanism providing both constraints and a wide range of variability at present is highly speculative. We offer it, not as a final answer but as a potential framework for scientific investigation. Undoubtedly, there will be a need for extensive refinements. After all, as stated in chapter 7, our understanding of regulatory mechanisms is still on the ground floor. Indeed, some would say that as yet we haven't even finished laying the foundation. But it is a start. Of course, we may be wrong. But that is the nature of science, and admittedly, as creationists, we have been far too timid to stick our necks out. It is time to bring forth constructive ideas boldly and openly—ideas that will lead to fruitful research. It is our sincere hope that this is one such idea.

The Created Kind

At the outset, many might complain that the concept of a ''created kind'' is not a legitimate scientific pursuit because its origin is from the Bible. The word *kind* is a biblical term that we do not hide from. As a matter of fact, many theories of science have had their origin from stranger sources. Kekule gained his inspiration for the ring structure of benzene from a dream of a snake biting its tail. Tesla got the idea for the alternating current motor from a vision while he was reading the poet Goethe. The point is that one's source for an idea is irrelevant. The key is whether the theory is practical and testable. If our theory of the created kind is to be rejected, let it be for a lack of scientific integrity and not because its inspiration is biblical.

To help alleviate any bias that may arise from the use of the phrase "created kind," we would like to propose the word *prototype* in its place. *Proto* being the Latin prefix meaning "ancestral" and *type* deriving from the Latin word *typus,* meaning "image." This provides a Latin base consistent with the rest of taxonomy while providing for the essential meaning of an ancestral image or form on which numerous variations are possible.

It is admittedly not an easy task to submit a definition of a prototype that has significant biological meaning. In taxonomy, this difficulty is not without precedent. There is no universal agreement over what a species is, let alone whether there are two, three, or five kingdoms. In evolutionary taxonomy, though, all organisms are said to be related through descent, the species being the only category that has any criteria that can be said to be reasonably objective. Unfortunately, one cannot always determine whether two populations are reproductively isolated or not. There are degrees of isolation that make even species designations tenuous. But, at least, in many cases reproductive isolation is a testable entity. Not so for categories from genus on up to kingdom. These distinctions are based primarily on degrees of similarity and dissimilarity. With the inclusion of the prototype, a new, frequently testable taxonomic category is introduced.

In the broad sense, by a prototype we mean "all organisms that are descended from a single created population." This definition necessitates making two distinctions. First, the prototype is not synonymous with species. To speak of the fixity of species is outdated and inaccurate. A prototype may consist of only one species or it may be composed of dozens of species. As mentioned previously, some groups, due to organismal and ecological characteristics, speciate more readily than others. This concept would hold true under the banner of prototypes as well. Second, the prototype cannot be universally associated with any particular taxonomic level. Evolutionists who demand identification of the created kind with some particular taxon are simply forgetting the subjectivity of higher taxonomic levels. In some groups, it may turn out to be synonymous with family, and in others, with genera or even species. The prototype is a discrete and potentially objective unit based on the assumption that all species within a prototype are descended from a single initial population. Therefore, it will not be useful to attempt broad identification with any one taxonomic unit across the board.

This makes the identification of the prototype more difficult. In fact, it may prove to be a monumental task for creationary biologists. The identification of the prototypes becomes a process of reevaluation according to traditional and nontraditional methods.

1. *Morphology*. Since Linnaeus's morphology has been the predominant method of distinguishing taxonomic categories. This would still make a useful initial step. A woodpecker is not an ostrich, and a rose is not a dandelion. The current taxonomic levels of kingdom, phylum, and class would still be most helpful at this point.

2. *Embryology*. Embryology may prove helpful in beginning to determine the role of regulatory mechanisms in embryonic development. These mechanisms could also be carried over into postnatal development. Fundamental differences here may lead to distinguishing characters in terms of the regulatory structure.

3. *Chromosome Morphology*. Since differences in gross chromosome morphology usually

are the cause of hybrid sterility, it follows that chromosome morphology may help in identifying members of a particular prototype, as well as excluding others. It is interesting to note that in 1977 a standardized karyotype was proposed for the North American deer mouse (*Peromyscus*), which includes more than fifty species.[174] With all species studied, the consistent element is chromosome number (2n = 48). Species differ in arm number or fundamental number (NF = 56-96). The standardization was made feasible because of the virtually identical G-banding patterns of the karyological extremes for the euchromatic arms. Surely, these are all related species within one prototype. The next step would be to compare chromosome morphology with those of other genera with similar organismal morphologies, such as the house mouse (*Mus*).

4. *Structural genes.* The use of structural genes involves not only amino-acid-sequence differences of protein, such as cytochrome *c*, common to most organisms, but also delineation of proteins unique to certain groups. This process is one area we know very little about at present, but it holds a great deal of potential, if we are applying information theory correctly.

5. *Regulatory mechanisms.* This is undoubtedly the critical area of investigation of the future. We know so little about it. We have a fairly good concept of the relationship between a single structural gene and its protein product. Our knowledge of the relationship between the gene and the organism as a whole is virtually nil.

> Nobody at this instance has the faintest notion, for example, how differences in genes make a difference in the shape of my nose or my ears or my behaviour or anything of this sort. There is a huge gap between genes on the one hand, and organism on the other.[175]

Much of this mystery undoubtedly is caught up in the regulatory mechanisms. But here is the key that we believe will set apart the prototype and define the limits to biological change. The above criteria are not new and have been used by taxonomists for decades. However, we believe that all members of a prototype will possess the same regulatory and developmental pathways. Since the "rules of grammar" do not originate by natural processes, the regulatory and developmental pathways will remain stable through speciation and mutation events. These pathways will differ from one prototype to the next.

A host of other biological facts could also serve as additional criteria. These would include behavior, physiology, reproductive patterns, and success or failure of gametic fusion. In short, in many cases, it will require a very extensive knowledge of the organism to make the identification.

The inclusion of the prototypes into the higher categories will continue to be helpful. The initial taxonomic categories by Linnaeus were based on creationist thinking; so they should pose no real problem. Cladistic taxonomy has already made a start in the direction of classifying on a basis of natural groups rather than phylogenetic relationships. For instance, though we propose that not all mammals are ancestrally related, there are a number of characteristics in common between prototypes that would allow them to be grouped together in the class Mammalia.

At times various levels of data may seem to conflict. An illustrative example is that of humans and the great apes. Anatomically and behaviorally, humans are quite distinct from the orangutan, gorilla, and chimpanzee. However, genic and highly detailed chromosome studies show an unexpectedly high degree of similarity. We have already commented in chapter 7 on the lack of certainty of the King and Wilson[176] conclusions based on genic data. However, it should be noted that regulatory differences would still be more than adequate justification to place humans and chimps in separate proto-types. The karyological data are more interesting.[177] All four species show remarkable similarity in G-banding patterns. A reversal of the numerous inversions (para- and pericentric) and the few fusions, insertions, and translocations would result in nearly 100 percent homology of banding patterns. This would be expected under the assumption that man and the apes are evolutionarily linked, yet contradictory to the concept that the four species (or at least humans) are members of discrete prototypes.

It must be noted, however, that though the similarities are great, the differences are not trivial. There are ten differences of chromosome arrangement between humans and chimpanzees. Nine pericentric inversions and one chromosome fusion separate the chromosomes of humans from those of chimps. The differences between humans and the gorilla and the orang-utan are even greater. By Bush's criteria,[178] humans and the great apes possess reproductive and ecological strategies unsuited to the establishment of chromosomal rearrangements. They have a low reproductive rate, late sexual maturity, few offspring, long life span, and high competitive ability. They are capable of wide-ranging migration patterns and are generalized feeders. These are characteristics of groups speciating by the classic allopatric model, which is very slow. Chromosomal rearrangements are difficult to fix because of partial hybrid sterility and, depending on the organism, substantial gene flow between populations. The sterility is due to the problems at chiasma. With pericentric inversions, some portion of the chromosome is duplicated and some of it is deleted. Greater difficulty persists, then, for hybrids of a centric fusion. Unbalanced gametes may result with either an extra chromosome or a whole one deleted. What is needed to fix such changes are small, isolated, fast-breeding, inbred populations. If humans and chimps split off only 5 million years ago, as some now say, these types of species have had a large number of chromosomal changes for such a short amount of time. Obviously, then, there are problems and questions on both sides.

But just as the evolutionist can fall back on long time spans and unusual speciation events or perhaps some unknown selective advantage, so the creationist can fall back on the commonality of design. But we suspect that neither answer by itself is fully satisfying to the inquisitive mind. This is one area of research that is of mutual and crucial interest to both evolutionists and creationists.

To summarize our concept of the prototype, let us emphasize that a host of biological techniques may be applicable in identifying the members of a kind, as well as distinguishing between the prototypes. No one criterion such as chromosome morphology, behavior, gametic fusion, etc., will be definitive in all cases. At times, a single test may be sufficient; at others, a multiplicity of approaches will be necessary. However, in terms of the mechanism of limited variation, the application of information theory to the genetic machinery should prove the most promising. The crucial factor will be delineation of the necessity of *intelligent* design in the structuring of the informational content and grammar of the genome of each prototype. This will indicate the necessity not only of intelligence in originating the genetic code in the broad universal sense but also, in the specific sense, of the unique adaptive programs of each prototype.

But just as a word of caution, it is necessary to point out that although we propose that the prototype has an objective reality, we may only be able to hypothesize about a particular organism's inclusion in a prototype. From our point of view, the prototype was an original stock with an objective existence. We may only hypothesize the descent of present-day creatures from particular prototypes. Our hypotheses will be founded on "scientific" observations, but they will for the most part remain hypotheses. The taxonomic ambiguity of the giant panda, then, would be a problem for the creationist as well as for the evolutionist.

Speciation and the Prototype

We would predict that speciation events would not affect the basic design of the organisms. This would simply be a splitting up of the gene pool of one species to create two. With the constraining factors of the regulatory mechanisms still operating and the inability of mutations of any type to produce new genetic information, the maintenance of the basic plan is to be expected. One need only recall the evidence of various rapidly speciating groups such as the Hawaiian *Drosophila*, the North American pocket gopher (*Geomyidae*) and the North American minnows (*Notropis*). In all these cases, the basic overall form is maintained. Even in such seemingly divergent island groups as the Galapagos finches and Hawaiian honeycreepers, the differences may be due to the combination of genetic drift, founder effect, recombination, and selection. Mutations need not play a role. Certainly within the context of information theory, there is a wide range of variability within a given vocabulary and set of grammatical rules.

A further result that we would expect is that speciation will often increase the chances of extinction to the newly formed species. This would be the result of diminishing the genetic variability because the speciating group contains a nonrepresentative sample of the gene pool. This would lead to a general prediction of susceptibility to extinction because of stochastic environmental events within rapidly speciating groups. Another contributing factor would be the subdividing of the available ecological space, thus creating smaller population sizes more susceptible to single-blow extinction. A case in point would be the desert pupfish of Death

Valley. The future of these relic species is not very bright, because of their low population levels and highly specialized habitat. It appears that they would have very little chance of surviving a major environmental shock, whether humanly or naturally induced.

Another observation we must make is that speciation may not necessarily be an adaptive event. It may simply result from a nonadaptive accident or fortuitous series of events. What adaptive divergence did take place would occur primarily after reproductive isolation was established. A good example is the pocket gopher, because in most cases one is hard pressed to find any adaptive difference between various chromosomal races and species. The same holds true for the Hawaiian *Drosophila* since all that separates the species at times is a difference in reproductive behavior. And this primarily serves as the reproductive isolating mechanism.

The Source and Meaning of Genetic Variation

Observable variation within a species and between species will have four possible sources: environment, recombination, mutation, and creation. Environmental effects are not heritable, though not necessarily unimportant. Phenotypic plasticity can be very crucial to the survival of the individual. A genotype is said to be phenotypically plastic (versatile) when its expression is variable according to the environment. The documented increase in human stature over the past few decades is probably due more to nutritional and hygenic factors than to any genetic change. The tanning process of human skin is another example. The tanned skin serves as a protective layer against further damage to the skin from ultraviolet rays. But such variation cannot be directly passed on to future generations. The *ability* to grow to seven feet or of the skin to tan may be passed on, but not the actual seven-foot height, nor the tanned skin. Therefore, although such plasticity can be important to individual survival, it has no lasting effect on the genotype. If it did, Lamarck would be vindicated.

A second and most important source of variation is recombination. This is best exemplified by the numerous examples of artificial selection that have already been mentioned. Within nature, the phenomenon of adaptive radiation is also explainable in terms of recombination combined with founder events. The Galapagos tortoises and finches, the Hawaiian honeycreepers, and the Lake Victoria cichlids can all be accounted for by segregation of variability into reproductively isolated groups taking advantage of diverse yet unoccupied habitats. This would fit in very well with our application of information theory. Variety is surprising, seemingly novel, given the same set of rules but with different subsets of an original vocabulary. Such would be the same with genetic information.

A third source of variation is mutation. Mutations, being mistakes in the genetic copying process, will be unable to add new genetic information to the genome. They will be capable only of modifying what already exists, usually in a meaningless or deleterious way. That is not to say that beneficial mutation is prohibited; unexpected maybe, but not impossible. A beneficial mutation is simply one that makes it possible for its possessors to contribute more offspring to future generations than do those creatures that lack the mutation. For example, in 1914 there occurred in Florida a mutation in the tomato that caused a change in growth pattern, making tomatoes much easier to harvest. This mutation, through human selection, has now spread throughout the cultivated tomato. The mutation for antibiotic resistance in bacteria is certainly beneficial for those bacteria whose environment is swamped by antibiotic. But these mutations have nothing to do with changing one kind of organism into another.

A type of change that is more significant

involves the loss or decrease of some structure or function. In this regard, Darwin called attention to the wingless beetles of Madeira. For a beetle living on a windy island, wings can be a definite disadvantage. Mutations causing the loss of flight are definitely beneficial. Similar would be the case of sightless cavefish. Eyes are quite vulnerable to injury, and a creature that lives in total darkness would benefit from mutations reducing their vulnerability. While these mutations produce a drastic and beneficial change, it is important to notice that they always involve loss, never gain. One never observes wings or eyes being produced in species that did not previously possess them.

Overall, however, mutations would primarily be a constant source of genetic noise and degeneration. It is unreasonable to expect natural selection to weed out all undesirable mutations that have occurred since the creation of each prototype. Total maintenance of the perfection of adaptation initially created is not expected. Many critics of creation have made such unreasonable claims, however. They argue that imperfection of design of adaptation argues forcefully for the reality of the evolutionary process. Evolution does not always find the best solution, just one that works.[179] It is said that imperfections are signs of history: if each kind had truly been created, adaptation would in all cases be perfect. This is an amazingly naive position.[180] Imperfections may indeed be signs of history, but perhaps they are more a sign of degeneration due to mutations. On a larger scale, these imperfections are signs of degeneration manifested as patchwork strategies to survive, not of slowly perfecting evolution. Although surprisingly few people seem to realize it, imperfection is an important part of creationist thinking. The same sort of engineering analyses that reveal the existence of origin by intelligent design, also allow one to identify subsequent disordering events, most of which in living systems are traceable to the effects of mutations.

A most interesting application of this involves human evolution. Arthur Custance, a Canadian anthropologist, has proposed a history of additive imperfections to explain the variation of fossil human skulls.[181] Custance shows that the variation in skull shape and size of three breeds of dogs is comparable to skull differences between gorilla, *Homo erectus*, and modern humans. One clearly sees that *Homo erectus* may be within the range of human skull variability. He also demonstrates that diet may radically affect skull shape, one type of diet providing a primitive looking skull. The point is that populations of *Homo erectus,* since their brain size is within the range of modern humans, may be fully human populations. The effects of diet and genetic deformities due to mutation and inbreeding produce skeletal remains that appear primitive. Rather than being a proto-human, it is possible that *Homo erectus* is fully human, but in a degenerative state. Evolutionists would not be likely to make such a prediction since their search is primarily for apelike ancestors, not humans with variable characteristics. This is a good example of a creationist perspective leading to research questions otherwise left unasked.

Mutations and Molecular Homology

One last area regarding mutations that deserves comment is molecular homology. This is an emerging science of comparing amino acid sequences to proteins to help establish evolutionary relationships and times of divergence. The most well-known phylogeny is that of cytochrome $c,$ which appears to agree very well with the accepted phylogeny. However, there are exceptions and procedural difficulties of interpretation. There are often large discrepancies between the protein phylogeny and the traditional one. In cytochrome c chickens are more closely related to penguins than to ducks and pigeons, turtles are closer to birds than to snakes (fellow reptiles), and people and monkeys diverge from the mammals before marsupial kangaroos separate from the rest of the mammals.[182] Another problem is that from the raw data alone, not one single phylogeny emerges, but several. The one that agrees most closely with the traditional phylogeny is *assumed* to be the most "correct." This hardly demonstrates independent confirmation of evolutionary relationships. The combining of several phylogenies from different proteins combines not only strengths but also weaknesses.[183] We can expect little more to be gained. One pair of researchers found the attempts to establish phylogenetic relationships from molecular homologies so fraught with difficulties and contradictions that they suggested the following:

> We can leave aside the question how proteins evolved (diachronic approach). Instead we ask what is common in the structure of functionally homologous proteins we find nowadays, i.e., we can focus our attention on the structure (synchronic approach) disregarding the genesis.[184]

This whole issue of homology, whether in organismal structures or protein sequences, brings up a key presuppositional difference between evolution and creation. Similarity in structure may indeed be indicative of ancestry, but it may also merely indicate a similarity of function. The distinction between homology and analogy is difficult to distinguish in any given case. The commonality of proteins in all organisms is just as logically attributable to common design based on common metabolic needs as to a single common ancestor. Although cytochrome *c* has thirty-five amino acid residues common to all organisms, the differences in the seventy others are not all meaningless mutation-generated noise. There are signs of functional differences in the cytochromes *c* of different species.[185] As creationists we would predict that the majority, if not all, of these sequence differences have some functional basis. The assumption behind many protein phylogenies and particularly molecular clocks (protein phylogenies formerly determined the times of divergence) is that these amino acid substitutions are due to the accumulation of neutral mutations at a relatively constant rate.[186] This is one area where creationist and evolutionist predictions definitely diverge. Since each organism will have slightly different needs, each cytochrome c may have been designed slightly differently, albeit fully recognizable as cytochrome c.

It is also important to note that some proteins are more variable than others. This is also best interpreted on a functional basis. The histones, for example, are virtually identical in all organisms studied. Only two changes in the 102 residues have been found in the H4 histone.[187] There would be little need for change in the proteins that, along with the DNA, are the substance of chromosomes. Cytochrome c, however, being involved in the energy-producing electron transport chain, may need slightly differing affinities from one created kind to another. Based on a creationist approach, a promising avenue of research into protein structural similarity would be to investigate functional differences, rather than pursuing the confirmation of ancestral relationships based on the regular accumulation of neutral mutations.

The fourth and final source of variation is creation. This is necessary simply because the first three sources of variation are inadequate to account for the diversity of life present on the earth: (1) environmental variation is nonheritable; (2) recombination fails without genetic variation present; (3) and mutation is incapable of producing significant amounts of meaningful genetic variants. An essential feature of creationary biology is the inclusion of considerable genetic variety in each prototype. A diversified vocabulary of genetic information had to be initially supplied. This is imperative in explaining the origin of horses, donkeys, and zebras from the same prototype; of lions, tigers, and leopards from the same prototype; of some 118 varieties of domestic dog—as well as jackals, wolves, and coyotes—from the same prototype. The chance processes of recombination and the more purposeful process of natural selection caused each prototype to subdivide into the vast array we now see.

According to creationary biology, natural selection is relegated to a conservative role. This is how it was initially categorized by the creationist Edward Blyth, twenty-four years before Darwin.[188] By the logical facts that some organisms will produce more offspring than others and that those that do are usually adequately adapted, deleterious mutations will be eliminated. The modern theory of environmental tracking is a helpful concept. Depending on the variation already present in the population, organisms can be expected, first of all, to survive and, second, to make minor adjustments of environmental changes over time. In the context of information theory and thermodynamics, natural selection is a cybernetic system. Cybernetics is the science of maintaining order in a system. Since all systems eventually show a tendency toward higher levels of entropy or disorder, their random deviations from order must be continually corrected. In living systems, these random fluctuations are mutations, which natural selection eliminates.[189]

Darwin's picture of natural selection as a creative force has prove unjustifiable. The criticisms of natural selection offered earlier were aimed at its hypothetical creative role. As we mentioned earlier, it is a real phenomenon, but it cannot explain the origin of new and often more complex structures. Natural selection, recombination, mutation, and speciation can all interact in concert to bring about startling variation within the created prototype without violating their integrity. These processes, along with the initial created variation are eminently qualified to account for the beauty, variety, complexity, and mystery in nature. This is not to say that all further research is superfluous because we have it all figured out. On the contrary, nature will provide an endless storehouse of unknowns as we search for ways to carry out the command to rule and have dominion (Genesis 1:26–28). To rule wisely and effectively, however, we must first comprehend. The Creator has revealed aspects of Himself through what has been made (Romans 1:18–29). Much, perhaps most, of what has been created remains unknown, yet discoverable. And we have been challenged to discover it. We can think of no greater motivation than this for engaging in scientific research.

The Controversy: A Call to Reason

It is unfortunate that in most cases, discussions involving the issue of creation and evolution tend to generate more heat than light. This is most often due to a lack of understanding of the basic presuppositions of the two polarized positions (more on that later). The other contributing factor is that the question of origins deals with history—i.e., unique events. One cannot study history in the same way that one can study gravity. Gravity is observable today; history leaves us only artifacts to piece together, much as the lawyer does in a courtroom. Both can be studied scientifically, one directly, the other indirectly. A test cannot be devised for evolution and creation as it can for gravity. Many evolutionists have complained about the statement by Karl Popper that Darwinism is not a scientific theory, but a metaphysical one.[190] He goes on to state that Darwinism is of great value to science as a metaphysical research program. The point is that evolution in the broad sense, from molecules to man, is not directly testable. That does not mean it is not scientific. One can *derive* predictions that in *many* cases can be tested.[191] Popper was not intimating that an evolutionist is not a scientist. Creation is in the same category. It is not directly testable, yet we can derive testable predictions, as we have attempted in this chapter. Some have declared that there are no scientists who are creationists. This is nothing more than schoolyard name-calling; it is simply not true. It is sad that such accusations have been deemed necessary by some.

Because both creationist and evolutionist approaches to origins are metaphysical constructs from which scientifically testable hypotheses are derived, there arises another sticky problem. A single falsified hypothesis on either side is not likely to cause a rejection of the complete metaphysical package. "We just asked the wrong question" is the most likely response. Or, some way will be found to work the new data into the model. Neo-Darwinism was not flatly rejected when significantly greater amounts of genic variation were found than were ever expected. The theory was merely adjusted to encompass the new data, though the neutralists still doubt whether any adjustments were necessary. Creationists, on the other hand, realized that they had formulated the wrong hypothesis when it was established that the species was not immutable. It will take a long series of defeats for adherents of either persuasion to change their minds. One consequence of this overall flexibility is that theorists from both sides are able to explain essentially all the relevant data within the confines of their own model. It is difficult, then to derive tests that effectively separate the two basic frameworks. As a result, evolutionists and creationists often compete by trying to come up with the "best" explanation for their own points of view. Consequently, each side attempts to make its own explanation more plausible by criticizing the other side's conclusions. Creationists have often been criticized for being overly negative. Although the major tenor of this book has been aimed at a critique of evolutionary mechanisms, we hope that this chapter has made an attempt to offset the imbalance. But when any view is in the minority (i.e., creation), the majority view (i.e., evolution) must be seen as sufficiently less than satisfying for alternatives to be realistically considered.

The other critical factor involves the presuppositions or world view we hold. The inability to recognize our own set of presuppositions can have devastating effects. David Bohm observes:

> It seems clear that everyone has got some kind of metaphysics, even if he thinks he hasn't got any. Indeed, the practical "hard-headed" individual who "only goes by what he sees" generally has a very dangerous kind of metaphysic, i.e., the kind of which he is unaware, . . . dangerous because, in it, assumptions and inferences are being mistaken for directly observed facts with the result that they are effectively riveted in an almost unchangeable way into the structure of thought.[192]

The single most basic presumption is the existence or nonexistence of God. If God exists, creation in some form is a possibility, though not the only one. If God does not exist, some form of evolution is the *only* possibility for origins. As all can readily see, our acceptance or rejection of the existence of God can have a profound effect on our thinking involving creation and evolution. Of course, some can believe in God and yet believe that He has no dealings with the material universe. Evolution again becomes the only choice. Scientific investigation into the subject of origins is not, therefore, done in some sort of ideological vacuum. Greene, in his book *Science, Ideology, and World View*, makes this very clear about science in general.

> Science, philosophy, theology, and other forms of rational inquiry are not totally insulated from the social, economic, psychological, and cultural con-

texts in which intellectual endeavor takes place. Even the sciences of nature can lay claim to no such intellectual purity. In the short run, science itself is shaped by existing knowledge and ideas. General conceptions of nature, God, knowledge, man, society, and history dictate what kind of science, if any, will be attempted, what methods will be employed what topics will be investigated, what kinds of results will be expected; social norms define the value and purpose of these inquiries. In the long run, however, if the scientist has insight and intellectual integrity, his findings may alter the general conceptions that shaped his inquiry despite his own reluctance to give up received ideas. Or he may adopt a radical stance and use his findings as ideological weapons, extrapolating science into world view. Like some modern evolutionary biologists, he may find himself writing a book entitled *The Meaning of Evolution* or *Evolution in Action* or *Nature and Man's Fate*. The lines between science, ideology, and world view are seldom tightly drawn.[193]

World view colors our thoughts in everything we do. Science is no exception, and the concept of origins is an integral part of one's world view. As a result, a scientific investigation into the subject of origins will be greatly affected by our world view, particularly in light of the difficulty in finding adequate tests, as we have seen. A knowledge of how we got here and why we are the way we are is crucial to understanding our world and how we react to each other. Whereas a theistic view of origins must necessarily affect the world view of its adherents, those who accept any form of evolution must realize that their position has no less of an effect on their world view. Greene points this out particularly in regard to Darwinism:

Darwinism, in various transmuted forms, is alive and tolerably vigorous in our own time, as in the writings of leading evolutionary biologists like Julian Huxley, George Gaylord Simpson, Theodosius Dobzhansky, and Edward O. Wilson. To the historian of ideas, these writings display the same interplay of science, ideology, and world view that characterized the words of Darwin and his contemporaries. They dispel, or at least should dispel, the dream of a purely scientific view of reality. Science is but a part, though an important one, of man's effort to understand himself, his culture, his universe.[194]

Gould and Eldredge have claimed that the gradualistic nature of Darwinism grew out of the cultural and political biases of nineteenth-century liberalism.[195] By contrast, it is clearly recognized by others that Gould's ideas of punctuated equilibria are guided by his own Marxist leanings.[196] Gould even admits to "surprise" on discovering that many Russian paleontologists adhere to a similar version of evolutionary change as he and Eldredge.[197] It is most unfortunate that the ideological underpinnings of debates among evolutionists are easily recognized and discussed, but when we turn to evolution versus creation, we are suddenly confronted with the "science" of evolution against the "religion" of creation.

Rupert Sheldrake, in his discussion of the origin of novel structures, offers a refreshing and sensible appraisal of the predicament:

> The origin of new forms could be ascribed either to the creative activity of an agency pervading and transcending nature; or to a creative impetus immanent in nature; or to blind and purposeless chance. But a choice between these metaphysical possibilities could never be made on the basis of any empirically testable scientific hypothesis. Therefore, from the point of view of natural science, the question of evolutionary creativity can only be left open.[198]

No matter what we might think of Sheldrake's theory of formative causation, this kind of clear logical thinking is sorely needed today.

Objective evidence is available in support of Neo-Darwinism, punctuated equilibria, and creation. Each individual must weigh and interpret the evidence through his own world view and decide which fits reality most consistently. At present, all three are defended by educated individuals trained in a broad spectrum of science. Although one may be more plausible to the reader than another, none can be shown to be totally implausible. There is a great deal of research to be done. Truth is not arrived at by a majority vote. It is perhaps instructive to note that all majority views were once held by a minority. And many majority views of the past are now obsolete.

Response

V. ELVING
ANDERSON

This book deals with an important topic. A description of the astounding amount of variety among biological organisms is a major topic in biology. The underlying problem, of course, is to try to account for the origin of this variability.

A further question deals with the nature of adaptation. Organisms are found to live in those kinds of environments for which their structure and function give them a close fit. In part this arises from their developmental physiology, so that organisms have the ability to migrate or otherwise adjust to varying environmental circumstances. But there are also the long-term questions: Do organisms change over many generations and, if so, how are these changes related to the environment in which they are found?

In addressing these questions, the field of genetics is becoming more and more important. The continuity from one generation to the next obviously involves the transmission of genetic material. With the advent of recombinant DNA methods, we are getting a much better view of genetic change over time. For example, it is now clear that genetic elements can move about from one part of a chromosome to another, and they can also move from one species to another.

The authors of this book have given a fair and open treatment of current thinking on these topics. Some writers, for example, have completely dismissed the idea of natural selection, but Lester and Bohlin have given a thoughtful review of this phenomenon. The references also are an excellent help for those who wish to explore a particular topic more extensively.

Clearly the authors are presenting the issue so that the reader will be encouraged to arrive at his

183

or her own conclusion. An important step in this direction is to understand how it is that different scientists appear to come to conclusions that are diametrically opposed. Such differences in interpretation may arise from differences in the questions asked, the assumptions made, or the data examined. The approach used in this book should help readers to sort out these issues.

A central point made by the authors is to question the usual assumption that all organisms are biologically related. Sometimes that assumption is advanced as a reason for not believing in God as creator. It is this arbitrary barrier to thinking that the authors are trying to explore.

This book is not presented to the readers as the final word on the topic. Our expanding knowledge of the nature of the genetic material, mainly as a result of the use of recombinant DNA techniques, means that any book will be somewhat out of date by the time it is published. I would urge that the authors consider a major revision of this work not less than five years from now.

I have some personal differences of opinion and of emphasis. For example, I am not convinced that the Genesis account clearly requires limits to changes. Also, I do not think that punctuationalism is as distinct from gradualism as the authors suggest. A recent review indicates that even Darwin was not a pure gradualist. It seems very likely that there may have been long periods of time with little change in a specific set of species, followed by episodes of more rapid change. Such differences, however, whether held by me as a reviewer or by the reader, do not diminish the value of a book as an exploration of the questions. This is a solid and thoughtful treatment of a particular creationist position.

I am not sure that the term "prototype" will turn out to be useful as an alternative of "kind." Prototype has other connotations in my mind that are not easily dislodged. Furthermore, although the authors are very sensitive to the need for a testable definition, I am not sure that they have arrived at that point. It is clear, however, that their discussion should advance the search for the best term and for appropriate ways in which to test the concept.

Earlier theological literature gave considerable space to an "argument from design." The illustrations that were used necessarily were based on the science of that time. Scientific advances understandably have rendered the earlier discussions somewhat obsolete. Thus it is not surprising to see the basic question returning in a new context. Attempts by some scientists to explain how DNA could rise from natural causes are a variant of this issue.

The existence of the universe and its complexity seem to be sufficient grounds for raising questions as to its origin. A thoughtful person must choose whether or not to believe in God as creator and sustainer. The possibility that evidence for limits to change provides significant leverage on this question is the point of this book and is a topic that deserves attention.

The message of this book, if I understand it correctly, is to keep an open mind. Look openly at varying points of view and try to identify the assumptions that are made. Do not, however, let unanswered questions block your honest approach to the underlying basic question of God's existence.

References

[1] S. E. Aw, *Chemical Evolution: An Examination of Current Ideas* (San Diego: Martin Books, CLP Publishers, 1982).

[2] J. Kerby Anderson, *Genetic Engineering* (Grand Rapids: Zondervan, 1982).

[3] J. L. Hubby and R. C. Lewontin, "A molecular approach to the study of genic heterozygosity in natural populations. I. The number of alleles at different loci in *Drosophila pseudoobscura*," *Genetics* 54 (1966): 577–94.

[4] T Dobzhansky, "Nothing in biology makes sense except in the light of evolution," *American Biology Teacher* 35 (1973): 125–29.

[5] T. Dobzhansky, *Genetics and the Origin of Species* (New York: Columbia University Press, 1937).

[6] J. S. Huxley, *Evolution: The Modern Synthesis* (New York: Harper, 1942).

[7] E. Mayr, *Systematics and the Origin of Species* (New York: Columbia University Press, 1942).

[8] G. G. Simpson, *Tempo and Mode in Evolution* (New York: Columbia University Press, 1944); and G. G. Simpson, *The Major Features of Evolution* (New York: Columbia University Press, 1953).

[9] G. L. Stebbins, *Variation and Evolution in Plants* (New York: Columbia University Press, 1950).

[10] V. A. McKusick, *Mendelian Inheritance in Man*, 4th ed. (Baltimore: Johns Hopkins Press, 1975).

[11] T. Dobzhansky, F. J. Ayala, G. L. Stebbins, and J. W. Valentine, *Evolution* (San Francisco: W. H. Freeman, 1977), pp. 72–73.

[12] F. J. Ayala, "The Mechanisms of Evolution," *Scientific American* 239 (1978): 59.

[13] R. C. Lewontin, *The Genetic Basis of Evolutionary Change* (New York: Columbia University Press, 1974), p. 92.

[14] C. L. Markert, J. B. Shaklee, and G. S. Whitt, "Evolution of a Gene," *Science* 189 (1975): 102–14.

[15]W. M. Fitch and E. Margoliash, "The Usefulness of Amino Acid and Nucleotide Sequences in Evolutionary Studies," *Evolutionary Biology* 4 (1970): 67–109.

[16]Dobzhansky et al., *Evolution*, p. 107.

[17]J. R. Powell, "Genetic Polymorphisms in Varied Environments," *Science* 174 (1971): 1035–36; J. F. MacDonald and F. J. Ayala, "Genetic Response to Environmental Heterogeneity," *Nature* 250 (1974): 572–74.

[18]A. R. Templeton, "Adaptation and the Integration of Evolutionary Forces," in *Perspectives on Evolution,* ed. R. Milkman (Sunderland, Mass.: Sinauer, 1982).

[19]M. Demerec, "Origin of Bacterial Resistance to Antibiotics," *J. Bacteriology* 56 (1948): 63–74.

[20]Dobzhansky et al., *Evolution*, pp. 121–22.

[21]B. Kettlewell, *The Evolution of Melanism* (New York: Clarendon, 1973).

[22]F. J. Ayala, "Genotype, Environment, and Population Numbers," *Science* 162 (1968): 1453–59.

[23]F. J. Ayala, "Evolution of Fitness v. Rate of Evolution of Irradiated Populations of *Drosophila*," *Proc. Nat. Acad. Sci.* 63 (1969): 790–93.

[24]T. Dobzhansky, *Genetics of the Evolutionary Process* (New York: Columbia University Press, 1970), p. 200.

[25]Ibid., p. 201.

[26]I. M. Lerner, *Heredity, Evolution, and Society* (San Francisco: Freeman, 1968).

[27]C. M. Woodworth, E. R. Leng, and R. W. Jugenheimer, "Fifty Generations of Selection for Protein and Oil in Corn," *Agronomy Journal* 44 (1952): 60–66.

[28]J. L. King and A. C. Wilson, "Evolution at Two Levels. Molecular Similarities and Biological Differences Between Humans and Chimpanzees," *Science* 188 (1975): 107–16.

[29]R. K. Selander, "Genic Variation in Natural Populations," in *Molecular Evolution,* ed. F. J. Ayala (Sunderland, Mass.: Sinauer, 1976).

[30]R. K. Selander, "Phylogeny," in *Perspectives on Evolution,* ed. R. Milkman (Sunderland, Mass.: Sinauer, 1982).

[31]T. Dobzhansky, F. J. Ayala, G. L. Stebbins, and J. W. Valentine, *Evolution* (San Francisco: Freeman, 1977), pp. 129–30.

32 Paul S. Moorhead and Martin M. Kaplan, *Mathematical Challenges to the Neo-Darwinian Interpretation of Evolution,* Wistar Symposium No. 5. (Philadelphia: Wistar Institute Press, 1967), p. 74A.

33 Fred Hoyle and N. A. Wickramasinghe, *Evolution From Space* (London: Dent, 1981).

34 E. Chain, *Responsibility and the Scientist in Modern Western Society* (London: Council of Christians and Jews, 1970), p. 1.

35 Pierre P. Grasse', *Evolution of Living Organisms* (New York: Academic, 1977), p. 87.

36 Ibid., p. 130.

37 Francis Hitching, *The Neck of the Giraffe: Where Darwin Went Wrong* (New Haven, Conn.: Ticknor and Fields, 1982), pp. 56–57.

38 Dobzhansky et al., *Evolution,* p. 65.

39 Markert, Shaklee, and Whitt, "Evolution of a Gene," p. 112.

40 G. L. Stebbins, *Darwin to DNA, Molecules to Humanity* (San Francisco: Freeman, 1982), p. 157.

41 W. A. Frazier, R. H. Angeletti, and R. A. Bradshaw, "Nerve Growth Factor and Insulin," *Science* 176 (1972): 482–88.

42 R. L. Hill et al., "The structure function, and evolution of α-lactalbumin," *Brookhaven Symp. Biol.* 21 (1969): 139–52.

43 Eric R. Pianka, *Evolutionary Ecology* (New York: Harper & Row, 1978), p. 9.

44 T. Bethell, "Darwin's Mistake," *Harpers Magazine,* February 1976, pp. 70–75; N. Macbeth, *Darwin Retried* (New York: Dell, 1971), pp. 40–81.

45 C. H. Waddington, discussion of paper by Dr. Eden, in *Mathematical Challenges to the Neo-Darwinian Interpretation of Evolution* (Philadelphia: Wistar Symposium), pp. 12–19.

46 S. J. Gould, "Darwin's Untimely Burial," in *Ever Since Darwin* (New York: Norton, 1977), p. 39.

47 D. E. Rosen, "Darwin's Demon," *Systematic Zoology* 27 (3) (1978): 370–73.

48 S. J. Gould, "The Problem of Perfection, or How Can a Clam Mount a Fish on Its Rear End?" in *Ever Since Darwin,* p. 104.

49 An excellent review of this problem and others like it can be found in Hitching, *The Neck of the Giraffe,* pp. 85–103.

[50]R. C. Lewontin, "Adaptation," *Scientific American* 239 (1978): 212–30, see especially p. 216.

[51]Ibid., p. 222.

[52]Ibid., p. 230.

[53]Ibid., p. 230.

[54]T. Watanabe, "Infective Heredity of Multiple Drug Resistance in Bacteria," *Bacteriological Reviews,* 27:87–115.

[55]Richard P. Novick, "Plasmids," *Scientific American* 243 (1980): 102–23.

[56]F. W. Plapp and J. S. Johnston, "Evidence for the Primary Role of Regulatory Gene Changes in the Evolution of Insecticide Resistance in the House Fly," in *The Evolutionary Significance of Insect Polymorphism,* ed. M. W. Stock and A. C. Bartlett (Moscow, Idaho: University of Idaho Press, 1982), pp. 65–75.

[57]Dobzhansky et al., *Evolution,* p. 122.

[58]H. B. D. Kettlewell, "Further Selection Experiments on Industrial Melanism in the Lepidoptera," *Heredity* 10:287–301.

[59]F. J. Ayala, "Genetic Variation in Natural Populations: Problem of Electrophoretically Cryptic Alleles," *Proc. Natl. Acad. Sci.* 79 (1982): 550–54.

[60]R. K. Selander, "Genic Variation in Natural Populations," in *Molecular Evolution,* ed. F. J. Ayala (Sunderland, Mass.: Sinauer, 1976), pp. 21–46.

[61]M. Kimura, "Evolutionary Rate at the Molecular Level," *Nature* 217 (1968): 624–26; J. L. King and T. H. Jukes, "Non-Darwinian Evolution," *Science* 164 (1969): 788–98; T. Ohta, "Mutational Pressure as the Main Cause of Molecular Evolution and Polymorphism," *Nature* 252 (1974): 351–54.

[62]G. L. Stebbins, "Modal Themes: A New Framework for Evolutionary Syntheses," in *Perspectives on Evolution,* ed. Roger Milkman (Sunderland, Mass.: Sinauer, 1982), pp. 1–14.

[63]R. Milkman, "Toward a Unified Selection Theory," in *Perspectives on Evolution* (Sunderland, Mass.: Sinauer, n.d.), pp. 105–18.

[64]S. M. Stanley, *Macroevolution: Pattern and Process* (San Francisco: Freeman, 1979).

[65]D. Kitts, "Paleontology and Evolutionary Theory," *Evolution* 28 (1974): 466.

[66]J. K. Anderson and H. G. Coffin, *Fossils in Focus* (Grand Rapids: Zondervan/Probe, 1977), p.16.

[67]D. Raup, "Conflicts Between Darwin and Paleontology," *Bulletin Field Museum of Natural History* 50 (Jan. 1979).

[68]S. Stanley, *Macroevolution*, p. 39.

[69]S. J. Gould, *The Panda's Thumb* (New York: Norton, 1980), pp. 181, 189.

[70]Hitching, *The Neck of the Giraffe*, p. 19.

[71]Stebbins, *Darwin to DNA*, p. 491.

[72]N. Eldredge and S. J. Gould, "Punctuated Equilibria: An Alternative to Phyletic Gradualism, in *Models in Paleobiology*," ed. T. J. M. Schopf (San Francisco: Freeman, Cooper, 1972), pp. 82–115.

[73]S. J. Gould, *The Panda's Thumb* (New York: Norton, 1980), p. 182.

[74]S. J. Gould, "The Meaning of Punctuated Equilibrium and Its Role in Validating a Hierarchical Approach to Macroevolution," in *Perspectives on Evolution*, ed. R. Milkman (Sunderland, Mass.: Sinauer, 1982), pp. 83–104.

[75]Ibid., p. 86

[76]S. M. Stanley, *Macroevolution: Pattern and Process* (San Francisco: Freeman, 1979), 332 pp.; idem, *The New Evolutionary Timetable: Fossils, Genes, and the Origin of Species* (New York: Basic Books, 1981), 222 pp.

[77]Stanley, *Macroevolution*, pp. 122–32.

[78]C. Delamare-Deboutteville and L. Botosaneanu, *Formes Primitives Vivantes* (Paris: Harmann), 232 pp.

[79]G. G. Simpson, *The Major Features of Evolution* (New York: Columbia University Press, 1953), p. 331.

[80]Stanley, *Evolutionary Timetable*, pp. 85–86.

[81]S. M. Stanley, "Fossil Data and the Pre-Cambrian–Cambrian Evolutionary Transition," *American Journal of Science* 276 (1976): 56–76.

[82]S. J. Gould, *Ever Since Darwin* (New York: Norton, 1977), pp. 126–33.

[83]Stanley, *Evolutionary Timetable,: Fl p. 89*.

[84]Ibid., p. 91.

[85]Ibid., p. 93.

[86]Stanley, *Macroevolution*, pp. 181–212.

[87]D. M. Raup and S. J. Gould, "Stochastic Simulation and Evolution of Morphology—Towards a Nomothetic Paleontology," *Systematic Zoology* 23 (1974): 305–22.

[88]Stanley, *Macroevolution*, pp. 184–85.

[89]Gould, "Punctuated Equilibrium," pp. 101–2.

[90]Gould, "The Meaning of Punctuated Equilibrium," p. 92.

[91]S. J. Gould, "Is a New and General Theory of Evolution Emerging?" *Paleobiology* 6 (1980): 119–30.

[92]Gould, *The Panda's Thumb*, pp. 204–13.

[93]E. Mayr, *Populations, Species, and Evolution* (Cambridge: Harvard University Press, 1970), p. 12.

[94]Ibid., pp. 55–68.

[95]M.J.D. White, *Modes of Speciation* (San Francisco: Freeman, 1978); G. L. Bush, "Modes of Animal Speciation," *Annual Review of Ecology and Systematics* 6 (1975): 339–64.

[96]Mayr, *Populations*.

[97]Ibid., p. 124.

[98]Ibid., p. 309.

[99]Stanley, *Macroevolution*, p. 41.

[100]Ibid., p. 41.

[101]Ibid., pp. 41–43; Stanley, *Evolutionary Timetable*, pp. 111–12.

[102]E. Mayr, *Animal Species and Evolution* (Cambridge: Harvard University Press, 1963), pp. 579–80; idem, *Populations*, pp. 346–49.

[103]Stanley, *Macroevolution*, pp 43–47; idem, *Evolutionary Timetable*, pp. 112–26.

[104]Gould, "The Meaning of Punctuated Equilibrium," pp. 88–90.

[105]Stanley, *Macroevolution*, pp. 145–48; Gould, "New and General Theory," pp. 123–24.

[106]Bush, "Modes of Animal Speciation"; G. L. Bush, S. M. Case, A. C. Wilson, and J. L. Patton, "Rapid Speciation and Chromosomal Evolution in Mammals," *Proceedings of the National Academy of Science* 74 (1977): 3942–46.

[107]H. L. Carson, "The Genetics of Speciation at the Diploid Level," *The American Naturalist* 109 (1975): 83–92; H. L. Carson, "Chromosomes and Species Formation," *Evolution* 32 (1978): 925–27.

[108]White, *Modes of Speciation*.

[109]A. C. Wilson, G. L. Bush, S. M. Case, and M. King, "Social Structuring of Mammalian Populations and Rate of Chromosomal Evolution," *Proceedings of the National Academy of Science* 72 (1975): 5061–65.

[110]Ibid.

[111] A. C. Wilson, "Gene Regulation in Evolution," in *Molecular Evolution,* ed. F. J. Ayala (Sunderland, Mass.: Sinauer, 1976), p. 225.

[112] G. L. Stebbins, "Modal Themes: A New Framework for Evolutionary Syntheses," in *Perspectives on Evolution,* ed. R. Milkman (Sunderland, Mass.: Sinauer, 1982), pp. 12–14.

[113] T. Dobzhansky, *Genetics of the Evolutionary Process* (New York: Columbia University Press, 1970), p. 34; Mayr, *Populations,* p. 183.

[114] R. J. Britten and E. H. Davidson, "Gene Regulation for Higher Cells: A Theory," *Science* 161 (1969): 529–40; "Repetitive and Non-Repetitive DNA Sequences and a Speculation on the Origins of Evolutionary Novelty," *Quarterly Review of Biology* 46 (1971): 111–33; "Organization, Transcription, and Regulation in the Animal Genome," *Quarterly Review of Biology* 48 (1973): 565–613.

[115] J. W. Valentine and C. A. Campbell, "Genetic Regulation and the Fossil Record," *American Scientist* 63 (1975): 673–80.

[116] Ibid., p. 674.

[117] M. C. King and A. C. Wilson, "Evolution at Two Levels in Humans and Chimpanzees," *Science* 188 (1975): 107–16; Wilson, "Gene Regulation"; A. C. Wilson, S. S. Carlson, and T. J. White, "Biochemical Evolution," *Annual Review of Biochemistry* 46 (1977): 573–639.

[118] Stanley, *Macroevolution,* pp. 157–59.

[119] G. Oster and P. Alberch, "Evolution and Bifurcation of Developmental Programs," *Evolution* 36 (1982): 444–59.

[120] S. J. Gould, *Ontogeny and Phylogeny* (Cambridge: Harvard University Press, 1977); P. Alberch, "Ontogenesis and Morphological Diversification," *American Zoologist* 20 (1980): 653–67.

[121] R. Goldschmidt, *The Material Basis of Evolution* (New Haven, Conn.: Yale University Press, 1940).

[122] S. J. Gould, "New and General Theory," p. 125.

[123] G. L. Stebbins and F. J. Ayala, "Is a New Evolutionary Synthesis Necessary?" *Science* 213 (1981): 968.

[124] W. M. Fitch, "The Challenges to Darwinism Since the Last Centennial and the Impact of Molecular Studies," *Evolution* 36 (1982): 1133–43.

[125]J. S. Levinton and C. M. Simon, "A Critique of the Punctuated Equilibria Model and Implications for the Detection of Speciation in the Fossil Record," *Systematic Zoology* 29 (1980): 137.

[126]Ibid., p. 136.

[127]T. J. M. Schopf, "Punctuated Equilibrium and Evolutionary Stasis," *Paleobiology* 7 (1981): 156–66.

[128]T. J. M. Schopf, "A Critical Assessment of Punctuated Equilibria, I. Duration of Taxa," *Evolution* 36 (1982): 1144–57.

[129]B. Charlesworth, R. Lande, and M. Slatkin, "A Neo-Darwinian Commentary on Macroevolution," *Evolution* 36 (1982): 476.

[130]G. L. Stebbins, "Perspectives in Evolutionary Theory," *Evolution* 36 (1982): 1109–18.

[131]R. C. Lewontin, *The Genetic Basis of Evolutionary Change* (New York: Columbia University Press, 1974), p. 159.

[132]Ibid., p. 159.

[133]E. Mayr, *Animal Species and Evolution* (Cambridge: Harvard University Press, 1973), p. 12.

[134]G. L. Bush, "What Do We Really Know About Speciation?" in *Perspectives on Evolution*, p. 119.

[135]Ibid., pp. 119–20.

[136]Stanley, *Macroevolution*, p. 41.

[137]Mayr, *Populations*, p. 183.

[138]A. R. Templeton, "The Genetic Architecture of Speciation," in *Mechanisms of Speciation*, ed. M. J. D. White and C. Barigozzi (New York: Plenum, 1982).

[139]E. Mayr, "Speciation and Macroevolution," *Evolution* 36 (1982): 1119–32; A. R. Templeton, "Modes of Speciation and Inferences Based on Genetic Distances," *Evolution* 35 (1981): 719–29.

[140]M. E. Douglas and J. C. Avise, "Speciation Rates and Morphological Divergence in Fishes: Tests of Gradual Versus Rectangular Modes of Evolutionary Change," *Evolution* 36 (1982): 224–32.

[141]Mayr, *Populations*, p. 294.

[142]White, *Modes of Speciation*.

[143]R. J. Baker, S. L. Williams, and J. C. Patton, "Chromosomal Variation in the Plains Pocket Gopher, *Geomys bursarius major*," *Journal of Mammalogy* 54 (1973): 765–69; R. L. Honeycutt and D. J. Schmidly, "Chromosomal and Morphological Variation in the Plains Pocket Gopher,

Geomys bursarius, in Texas and Adjacent States," *Occasional Papers of the Museum, Texas Tech University* 38 (1979): 1–54; E. Nevo, Y. J. Kim, C. R. Shaw, and C. S. Thaeler, Jr., "Genic Variation, Selection, and Speciation in *Thomomys talpoides* Pocket Gophers," *Evolution* 28 (1974): 1–23.

[144]R. G. Bohlin and E. G. Zimmerman, "Genic Differentiation of Two Chromosome Races of the *Geomys bursarius* Complex," *Journal of Mammalogy* 63 (1982): 218–28.

[145]Bush, "Modes of Animal Speciation."

[146]Carson, "The Genetics of Speciation at the Diploid Level."

[147]Stebbins, "Perspectives in Evolutionary Theory," p. 1111; Charlesworth et al., "A Neo-Darwinian Commentary on Macroevolution," pp. 484–85.

[148]Stebbins and Ayala, "Is a New Evolutionary Synthesis Necessary?" p. 969; Charlesworth et al., "A Neo-Darwinian Commentary on Macroevolution," pp. 487–90; Mayr, "Speciation and Macroevolution," pp. 1127–29; Stebbins, "Perspectives on Evolutionary Theory," p. 1111.

[149]R. A. Fisher, *The Genetical Theory of Natural Selection* (Oxford: Oxford University Press, n.d.).

[150]K. F. Koopman, "Advanced Text on Macroevolution," *Bioscience* 31 (1981): 170.

[151]Charlesworth et al., "A Neo-Darwinian Commentary on Macroevolution," p. 486.

[152]Ibid., p. 488.

[153]P. Alberch, "Ontogenesis and Morphological Diversification," *American Zoologist* 20 (1980): 653–67.

[154]Stanley, *Evolutionary Timetable*, p. 119.

[155]Leigh Van Valen, "A Natural Model for the Origin of Some Higher Taxa," *Journal of Herpetology* 8 (1974): 109–21.

[156]Mark Ridley, "Who Doubts Evolution?" *New Scientist* 89 (25 June 1981): 830–32.

[157]Stanley, *Macroevolution*, p. 2.

[158]S. J. Gould, "The Validation of Continental Drift," in *Ever Since Darwin* (New York: Norton, 1977), pp. 160–67.

[159]Stanley, *Macroevolution*, p. 6.

[160]J. K. Anderson and H. G. Coffin, *Fossils in Focus* (Grand Rapids: Zondervan, 1977).

[161]Stanley, *Evolutionary Timetable*, p. 93.

[162]Lewontin, "Adaptation," *Scientific American* 239 (September 1978): 213.

[163]James E. Harigan, *Chance or Design* (New York: Philosophical Library, 1979), 233 pp.

[164]Carl Sagan, *Cosmos* (New York: Random, 1980), pp. 111–12.

[165]A. E. Wilder-Smith, *The Creation of Life* (Wheaton, Ill.: Harold Shaw, 1970), pp. 63, 239–44.

[166]Frank L. Marsh, *Variation and Fixity in Nature* (Mountain View, Calif.: Pacific, 1976), 150 pp.

[167]A. J. Jones, "The Genetic Integrity of the 'Kinds' (Baramins): A Working Hypothesis," *Creation Research Society Quarterly* 19 (1982): 13–18.

[168]A. J. Jones, "A Creationist Critique of Homology," *Creation Research Society Quarterly* 19 (1982): 156–75.

[169]Wilder-Smith, *The Creation of Life;* idem, *The Natural Sciences Know Nothing of Evolution* (San Diego: CLP, 1981); H. P. Yockey, "A Calculation of the Probability of Spontaneous Biogenesis by Information Theory," *Journal of Theoretical Biology* 67 (1977): 377–98.

[170]William R. Bennett, Jr., *Scientific and Engineering Problem Solving With the Computer* (Englewood Cliffs, N.J.: Prentice Hall, 1976), referenced in Jeremy Campbell, *Grammatical Man* (New York: Simon & Schuster, 1982), pp. 115–16.

[171]Campbell, *Grammatical Man,* pp. 147–58.

[172]Ibid., p. 149.

[173]Ibid., pp. 115–18.

[174]The Committee for Standardization of Chromosomes of *Peromyscus,* "Standardized Karyotype of Deer Mice, *Peromyscus* (Rodentia)," *Cytogenetics: Cell Genetics* 19 (1977): 38–43.

[175]Richard Lewontin, "Did Darwin Get It Wrong," *NOVA* #816 (Boston: WGBA Transcripts, 1981), p. 14.

[176]King and Wilson, "Evolution at Two Levels."

[177]Jorge J. Yunis and Om Prakash, "The Origin of Man: A Chromosomal Pictorial Legacy," *Science* 215 (19 March 1982): 1525–30.

[178]Bush, "Modes of Animal Speciation."

[179]Gould, *The Panda's Thumb,* pp.19–44.

[180]R. G. Bohlin and J. K. Anderson, "The Straw God of Stephen Gould," *Journal of American Scientific Affiliation* 35 (1983): 42–44.

181 Arthur C. Custance, *Genesis and Early Man* (Grand Rapids: Zondervan, 1975).

182 F. J. Ayala, "The Mechanisms of Evolution," *Scientific American* 239 (1978): p. 56.

183 C. Knaub and G. Parker, "Molecular Evolution?" *Impact,* no. 114, Institute for Creation Research, 1982.

184 A. Krzywicki and P. O. Slonimski, *Journal of Theoretical Biology* 21 (1968): 306.

185 S. Ferguson-Miller, D. L. Brautigan, and E. Margoliash, *Journal of Biological Chemistry* 251 (1976): 1104.

186 W. M. Fitch, "Molecular Evolutionary Clocks," in *Molecular Evolution,* ed. F. J. Ayala (Sunderland, Mass.: Sinauer, 1976), pp. 160–78

187 S. E. Aw, *Chemical Evolution: An Examination of Current Ideas* (San Diego, Calif.: Martin Books, CLP, 1982), p. 109.

188 Loren Eisley, *Darwin and the Mysterious Mr. X* (New York: Dutton, 1979).

189 Campbell, *Grammatical Man,* p. 23.

190 K. Popper, *The Unended Quest,* 1976.

191 K. Popper, "Letters," *New Scientist* 98 (21 August 1980): 611.

192 D. Bohm, "Some Remarks on the Notion of Order," in *Towards a Theoretical Biology,* vol. 2, ed. C. H. Waddington (Edinburgh: Edinburgh University Press), p. 41.

193 John C. Greene, *Science, Ideology, and World View* (Berkeley, Calif.: University of California Press, 1981), p. 2.

194 Ibid., p.8.

195 Gould and Eldredge, "Tempo and Mode of Evolution Reconsidered," p. 115.

196 Michael Ruse, "Charles Darwin and the Beagle," *The Wilson Quarterly* 6 (Winter 1983): 164–75.

197 Gould, *The Panda's Thumb,* pp. 184–85.

198 Rupert Sheldrake, *A New Science of Life* (Los Angeles: J. P. Tarcher, 1982), p. 150.

For Further Reading

NEO-DARWINISM

Dobzhansky, Theodosius; Ayala, Francisco J.; Stebbins, G. Ledyard; and Valentine, James W. *Evolution.* San Francisco: Freeman, 1977.

A detailed and technical description of Neo-Darwinian evolution. The topic of evolution is discussed from a diversity of perspectives such as zoology, botany, anthropology, and paleontology; physiology, microbiology, and biochemistry; population biology, ecology, and systematics; and genetics and developmental biology.

Stebbins, G. Ledyard. *Darwin to DNA, Molecules to Humanity.* San Francisco: Freeman, 1982.

Stebbins, one of the early developers of the modern synthesis, assesses the current status of Darwinism with a special emphasis on human cultural and biological evolution. This book is for the general reader.

Ruse, Michael. *Darwinism Defended.* Reading, Mass.: Addison-Wesley, 1982.

Although the subtitle of this book claims that it is a guide to the evolution controversies, it is exclusively a defense of classical Neo-Darwinism against all critics, particularly creationists and punctuationalists.

Grassé , Pierre P. *Evolution of Living Organisms.* New York: Academic, 1977.

Grassé totally rejects Neo-Darwinian explanations for the mechanism of evolution, though he is a convinced evolutionist. Grasse' is without question the leading French biologist of this century. A book for the more technical reader.

Macbeth, Norman. *Darwin Retried.* New York: Dell, 1971.

A logical critique of Darwinism by a lawyer.

Hitching, Francis. *The Neck of the Giraffe: Where Darwin Went Wrong.* New York: Ticknor and Fields, 1982.

An excellent summary of the logical and evidential pitfalls of Neo-Darwinism.

PUNCTUATED EQUILIBRIUM

Stanley, Steven M. *Macroevolution: Pattern and Process.* San Francisco: Freeman, 1979.

The only major work that systematically defines the basic aspects of punctuated evolution in technical detail.

Stanley, Steven M. *The New Evolutionary Timetable: Fossils, Genes, and the Origin of Species.* New York: Basic Books, 1981.

Essentially the same as the above book, except shorter and geared for the general reader.

Gould, Stephen J. *The Panda's Thumb: More Reflections in Natural History.* New York: Norton, 1980. See chapters 17–20.

Four individual essays by Gould that fluently discuss principles of punctuated equilibrium.

Evolution. Vol. 36, nos. 3 (May 1982) and 4 (December 1982).

These two issues of the journal Evolution *contain several articles that critique punctuational concepts in technical language.*

CREATION

Coppedge, James F. *Evolution: Possible or Impossible?* Grand Rapids: Zondervan, 1973.

A nontechnical discussion of the application of molecular biology and probability to evolution.

Wysong, R. L. *The Creation-Evolution Controversy*. Midland, Mich.: Inquiry, 1976.

A rational discussion of the implications and evidence regarding evolution and creation.

Wilder-Smith, A. E. *The Natural Sciences Know Nothing of Evolution*. San Diego: Master Books, 1981.

This is a detailed discussion of the origin and transformation of DNA in light of information theory.

Parker, Gary E. *Creation: The Facts of Life*. San Diego: CLP Publishers, 1981.

A nontechnical discussion regarding the evidence for creation principally from biology.

WORLD VIEW

Sire, James W. *The Universe Next Door: A Basic World View Catalogue*. Downers Grove, Ill.: Inter-Varsity, 1976.

An indispensable guide to what a world view is and the propositions of today's dominant world views.

Green, John C. *Science, Ideology, and World Views: Essays on the History of Evolutionary Ideas*. Berkeley: University of California Press, 1981.

These essays spell out clearly that science is not immune from the effects of one's world view. The author also spells out the dangers when this interplay is ignored.

Index

RNA (see Ribonucleic acid)
RNA polymerase 45, 47–48
Ruminant mammals 28

Sagan, Carl 154
Schopf, T. J. M. 133
Schutzenberger, Marcel 85
Sea cucumbers 133
Seastars 133
Selection
    artificial 78–79, 95–96, 170; balancing 70–71, 75, 104; differential 117; directional 70, 72, 75; diversifying 71; natural 66–67, 70–77, 93–96, 99–103, 108, 116–19, 121, 168, 175; normalizing 70–71, 94; species 116–19, 125; stabilizing 70–71, 133
Shakespeare, William 157–58, 160
Shekdrake, Rupert 135
Shrimp, Pederson 31
Simpson, George G. 66, 179
Skulls, human fossil 172
Speciation 73, 111–13, 116–22, 126, 132–38, 146, 165, 167, 175
    chromosomal 126; directed 116–17; and the prototype 168–69; quantum 128; rapid 122–26, 134, 136, 139, 142–44
Species
    biological 132; definition 119–20; paleontological 132; selection 116–19, 125; within prototypes, 162–63
Spencer, Hebert 93
Spider 27
Stanhopea grandiflora 29
Stanley, Steven 106, 108, 114–16, 123, 125, 128, 139, 142–44, 148
Stasis 106–7, 112–13, 130, 132–34, 150
Stebbins, George L. 66, 109, 127

Stochastic events 80–81
Sunfish 136
Supernumerary limbs 144–45
Survival of the fittest 93
Symbiosis 28–31
Sympatric 122
Synthetic theory 66–67

Tautology 93, 132, 146
Taxonomy 162–65, 168
Termite 25–26, 28
Tesla 161
Thomomys 137
Threonine 58
Thymine 34–40
Tomato 170
Tongue, woodpecker 24–25, 96
Tortoise, Galapagos 15–17, 170
Transcription 42–45, 47–48, 51, 103, 155
Translation 42–43, 47–48, 155
Translocation 59
Transposition 59
Triplet 40
Tryptophan 45, 58

Uracil 42, 57–58

Valentine, J. W. 127
Valine 58
Van Valen, Leigh 144–45
Vulture, turkey 28

Waddington, C. H. 93
Wasps 27
Whale 143–44, 150
White, M. J. D. 126
Wickramasinghe, Chandra 85
Wilson, A. C. 79, 128, 139, 165, 167
Woodpecker 23–25, 96–97

Zebra 100–101
Zygote 52

207

Interesting points

Ayala on randomness of mutations - 6
mutations occur in well
adapted individuals.